ABOUT THIS SERIES . . .

The ultimate responsibility for determining the fate of this nation rests with its citizens. Given that responsibility, citizens must inform themselves so they can make wise decisions. For the citizenry to do so requires, however, that sources of the facts and knowledge essential to the decision-making process must share in the responsibility by making information readily accessible to the public.

Because of the complexity and seriousness of the problems facing our nation today, special public communications efforts are needed more than ever before. This series represents one such effort by a basic industry, the nation's electric utility companies, through the Edison Electric Institute.

"Decisionmakers Bookshelf" seeks to provide to the public important discussions and reasoned viewpoints on national policy problems related to energy.

ABOUT EEI . . .

Edison Electric Institute is the association of America's investor-owned electric utility companies. Organized in 1933 and incorporated in 1970, EEI provides a principal forum where electric utility people exchange information on developments in their business, and maintain liaison between the industry and the Federal Government. Its officers act as spokesmen for investor-owned electric utility companies on subjects of national interest.

Since 1933, EEI has been a strong, continuous stimulant to the art of making electricity. A basic objective is the "advancement in the public service of the art of producing, transmitting and distributing electricity and the promotion of scientific research in such field." EEI ascertains factual information, data and statistics relating to the electric industry, and makes them available to member companies, the public, and government representatives.

A Short Energy History of the United States
(And Some Thoughts About the Future)

By Joseph M. Dukert

1111 19th Street, N.W.
Washington, D.C. 20036

Printed in the United States of America
at Sheridan Printing Company, Inc., Alpha, NJ

ISBN 0-931032-07-5

CONTENTS

INTRODUCTION

Considering the intense focus on the topic of energy in the United States since the early 1970s, the lack of integrated historical analysis is surprising.

Even though decisions about energy are made more carefully in this country today than ever before, history suggests they will produce some welcome results and some new problems. The threads of economic, social and political development have produced some discernible patterns over the years. There has been a general repetition in the way transitions have occurred; yet, the fabric also displays occasional unanticipated snags.

Citizens deserve to know what consequences one course or another is likely to produce, a judgment in which historical analysis can assist. Moreover, history should provide energy policy-makers with clues as to which of many interrelated factors might respond readily to actions being considered and which might not.

The author of this volume in Edison Electric Institute's Decision-makers Bookshelf series, Joseph M. Dukert, is a Washington, D.C. consultant who has been writing about energy for more than 20 years. He first became absorbed in historical relationships among various energy sources during a Bicentennial project for the Energy Research and Development Administration in 1975. The statistical information and conclusions presented here are based upon his own research.

I. THE "WHY" AND "HOW" OF THIS BOOK

"Energy" is a big word.

The term "energy" (as commonly used in today's headlines) really involves a relatively new concept—and a broad one. George Washington knew what "heat" was; it was one of the items in short supply at Valley Forge. To the boyhood Lincoln, "light" included the flicker of a log fire. The Adams family of New England might have pointed to water-driven textile mills as the example of "mechanical power" most familiar to them—even though a modern scholar could argue that the wind filling the sails of Yankee trading vessels in the early 1800s was at least equally important. But one of the characteristics of our *current* use of the word "energy" is that it can embrace *all* of those—and a good deal more.

Histories about individual energy sources exist. A student or policy maker has little trouble finding books and articles dealing with the development of coal . . . or oil . . . or nuclear power . . . or fuel gas . . . or solar energy . . . or wind machines. It is more difficult to find a history that addresses the interrelationship of all these, but even such a broader account might not be a true "*energy history.*" The pattern so far has been merely to describe "the history of energy"—a chronicle of technology which may intrigue the trivia fan but is of limited value to those who worry about today's energy dilemmas and clues to their possible solutions (including clues from the past).

Although historians haven't treated it as such, energy was as much of a critical factor in our "winning of the West," "industrialization," and later "urban decay" as it now is in U.S. diplomacy vis-a-vis the Middle East. But that role doesn't necessarily emerge from a mere chronology of energy-related inventions. In spite of its brevity, this book tries to capsulize the "energy history" *of our country* and to suggest some possible cause-and-effect relationships between patterns of energy use and various social developments. It looks at energy as a part of economic history; and, because energy has affected (and has been affected by) American politics and culture as well as science, it notes some of those links too.

History is a moving picture. We can stop the reel and examine it—one frame at a time; but much of its dynamism eludes us unless we analyze the historical record in time-chunks that are long enough to show trends. For that reason, this study divides the period between our Declaration of Independence and the shock of the Arab Oil Embargo into six segments of about 35 years each. Assuming that the U.S. turned another corner (or went over a hill) in 1974, we can speculate about what sort of energy phase our country may be in right now . . . and where it might lead us by the first decade of the 21st century.

The time-periods selected are unique to this book, and readers are welcome to quibble with them. Each period seems to present a distinctive character of its own, although some overlapping is inevitable. History is not always neat. Because space limitations permit only the inclusion of highlights here, it would be easy to point out certain exceptions and countertrends all along the line. Any critic who takes inordinate delight in doing so will have missed a basic point of this study: The energy problems we wrestle with today defy simple solutions *because* energy is so intertwined with so many diverse aspects of American life. It has been that way from the very beginning.

The principal statistical basis for this study (not only in energy production and consumption, but also in the socio-economic indices affected by them) was the U.S. Census Bureau's comprehensive *Historical Statistics of the United States: Colonial Times to 1970.* The most valuable single secondary source was the classic *Energy in the American Economy, 1850-1975*—compiled by Schurr, Netschert, Eliasberg, Lerner and Landsberg at Resources for the Future 20 years ago, but still fresh in its approach and insights. More recent data have come from various governmental and industrial publications.

Contemporary attitudes toward changing energy patterns over the whole span have been deduced partly from a survey of published anthologies (such as the multi-volume *Annals of America*) and partly from a great variety of secondary sources whose reliability could be verified. Because ideas have been drawn from all of these selectively, the validity of this entire presentation relies heavily on the author's success in empathizing with successive generations of Americans toward what we now recognize as critical choices regarding energy resources and their application.

Library shelves are already loaded with social histories of the United States—many of them instructive to today's energy planners. Nevertheless, whether such works cover centuries or shorter periods, they tend to treat so many other topics that it is difficult to find in them any continuous thread relating to energy. More surprisingly, the same applies to most histories of U.S. technology; their focus tends to be on things that *use* energy rather than on energy itself. Even the histories of individual energy sources and forms (coal, oil, electricity, solar, steam, etc.) don't tell a sufficiently rounded story to be as useful as we might like them to be in analyzing America's energy problems during the final decades of the 20th century. Our difficulties (and the answers to them) are not tied to either a single source or a single form. Edison Electric Institute, in financing this publication, laid down no conditions about the emphasis to be given to competing modes of meeting U.S. energy needs—in the past, present or future. The conclusions here are the author's.

Because U.S. energy history has not existed as a separate discipline, some popular myths about our past have generally gone unchallenged. For example, the 60-year cycles which some government policy documents have cited as reflecting our national shift from one fuel-form to the next (first from wood to coal, and then from coal to petroleum) are simply statistical coincidences. They have less than nothing to do with individual human lifestyles. Coal production did not soar during the second half of the 19th century in order to *replace* wood in its primary energy role of home heating; indeed, wood remained the basic residential heating fuel throughout most of the area of the Old South as late as 1950. The real story of how and why energy transitions occurred is far more intricate.

II. ENERGY HISTORY—THE FUTURE BEHIND US

Addressing a group of energy economists in Washington during mid-1979, then-Secretary of Energy James R. Schlesinger pooh-poohed the relevance of one mainstay of discussions about energy history—the "whale oil story." That story concerns how the rising price of whale oil in the mid-19th Century instigated a search for alternatives . . . and finally helped to launch petroleum-based kerosene as a substitute lamp fuel.

Schlesinger suggested that this particular analogy with modern times had been overdrawn, and he referred to those who "sell whale oil" as current-day "snake oil salesmen." Although the Secretary's cynicism had some justification (inasmuch as appeals to historic parallels can become too simplistic), those who deliberately ignore lessons from the past are teasing fate. Some patterns of history *do* repeat themselves. The difficulty (and the source of many arguments about energy policy for the future) lies in discerning on which ones to count.

Unsuccessful generals are frequently accused of trying to fight today's battles with yesterday's tactics. The United States would be foolish to pattern its responses for the 1980s on what it did (or should have done) in the 1880s. We are a different country now, living in a different world. Furthermore, one of the clearest lessons that develops from a careful review of the past is that we will be surprised again in the future. Benjamin Franklin, experimenting with electricity, did not actually foresee the incandescent light bulb—much less neon signs or fluorescent fixtures. But Franklin might have been less astonished by those inventions than by the thought that an anti-Federalist like Jefferson would take an expansionist-yet-centralizing step like the Louisiana Purchase. Jefferson's action was as crucial to the development of energy use systems in this country as any invention. It started a national move across the continent that would ultimately cry out for a transportation system like the railroad. It gave the national government control over vast resources (land and minerals) which could help to bring the rail network into being. It offered incentives—both in the New World and in the Old—for successive waves of immigration. It introduced a unique combination of private initiative and governmental stewardship that might involve continuing tension but would also produce unprecedented material progress. For all of that, the very opportunity to purchase the Territory of Louisiana came as a surprise, even to President Jefferson himself. And such is the stuff of energy history.

Once we accept the fact that we will not be able to predict *everything* in America's energy future, we can take heart from the simultaneous truth that there are threads of continuity. Certain basic relationships we can see throughout our past are likely to provide *some* warnings about what *may* happen in the years and decades to come.

The development of energy in the U.S. has been dispersed geographically, it has been largely spontaneous, and it has been pluralistic. Even if we limit the applications of energy to heat, light and motive power, there has never been a period in our history when we were able to get along on a single source. There is no reason to expect that to change. Wood was not our only energy source at the time of the American Revolution (although it was certainly an important one), and anybody who implies that the much more complex civilization we have built since then can survive by concentrating almost exclusively on a single contemporary source—be it coal, solar energy, nuclear fission, or anything else—is dangerously closed-minded about both energy history and the nature of energy itself.

There were obvious differences in the energy habits of diverse regions from the very beginning, and this is another characteristic which is likely to continue. Part of the reason is that resources are not distributed evenly; the early concentration on waterpower in New England and on use of natural gas in the Gulf states mirrors that. Another factor lies in regional differences among the sectors of energy consumption—agricultural, industrial, commercial, residential, and so on.

Perhaps the most important thing to note is that energy patterns in the past have responded most readily to the collective desires of many, many individuals. Price has played a role in popularizing or restricting the growth of new energy forms and energy applications, but has not been the only element in those individual decisions. Most Americans have always lived well above the subsistence level, and they have usually been willing to pay a premium for what they recognized as additional convenience, safety or comfort. At times we have made sacrifices (e.g., as in the case of supporters of the Confederacy during the Civil War, or of a good many Americans in World War II), but even then the unwelcome changes in energy habits were accepted on the basis of perceived self-interest.

Because of our individualistic approach, there have always been disagreements about the advisability of energy change. Over the years there have also been continual shifts in public mood. Environmental concerns are one example; there are contemporary accounts from almost every period that those were "the best of times" *and* "the worst of times." As late as 1880, municipal ordinances in Baltimore prevented Baltimore & Ohio Railroad trains from using steam locomotives as they moved through the area adjacent to the city's inner harbor, so teams of horses or oxen were substituted for them to draw the cars from one station to another. Later, the B & O was allowed to use locomotives—but only with the proviso that they be disguised to resemble electric streetcars which the city fathers believed were less likely to upset the thousands of horses pulling buggies, carts and wagons through the same streets. Apparently, those who disapproved of the noise, odor and general commotion caused by rail engines within the downtown area considered horse droppings environmentally acceptable.

Steam had real safety problems, however—like every other energy form and every other energy technology. The fact that occasional boiler explosions (or collisions, or derailments, or even coal-mine disasters) did not write an early end to the saga of the iron horse cannot be explained entirely by either public indifference or a muted expression of the public will. The American people, through their elected representatives, merely determined that benefits exceeded costs.

Relatively little of the nation's energy history has been moulded by express formulations of energy policy, even though statutes and regulations have always had an impact. Tariffs and taxation helped to determine the sorts of energy that would be needed and how they would be developed and applied. Social welfare legislation and labor law helped establish an underlying tone. The respective roles of free enterprise, consumers, and government at the local, state and national level have shifted continuously for over two centuries. They will continue to adjust to one another. They will also continue to affect the way we use energy—whether or not the word "energy" appears in the legal documents that are issued.

This country now uses more energy than ever before in its history, but the rate of growth in consumption has not been uniform. This is more than an intuitive assumption. It can be documented. But there are various ways of making the comparisons, and any technique is subject to disagreement. The author has tried to make the one used here as fair as possible, incorporating some factors which have often been overlooked.

First of all, there are some very basic energy sources that are not counted in the computations at all—even here—because they are so universally available and applied. The light and warmth we get from the sun is one example. Nutritional energy stored in natural foods is another. Gravitational energy, in the limited sense that we derive a free energy "bonus" by being able to roll things downhill, is still a third. The energy of moving water that drives a mill or a hydrogenerator, on the other hand, *is* counted. So is the energy derived from wind to propel a sailing ship or to pump water. By the same token, solar energy that is collected artificially by rooftop assemblies in order to provide heating or cooling inside ought to be included in the future as soon as enough devices are constructed to be noticed within the enormous nationwide statistics on all energy use. But so-called "passive solar energy" (e.g., "saved energy" that results from improving a house's insulation or planning its window orientation carefully) would not be included. That is a form of energy *conservation*, rather than true energy *supply*, and—for all its potential importance, some statisticians only confuse matters by switching such items from one side of the supply-demand equation to the other.

Various fuels, such as wood, coal and oil, can be compared with one another by considering their respective heat content, as in Figure 1. In this sort of comparison, for example, a ton of high-grade bituminous coal is equivalent to slightly more than four barrels (about 170 gallons) of heating oil.

Similar conversion equivalents exist for the mechanical work performed by wind and waterpower. In the case of hydroelectricity and nuclear power, the equivalency is based on the amount of fossil fuel that would be required to operate the steam boiler in a comparable generating plant.

The significance of increasing population in the United States' growing consumption of energy resources over the years has somehow been overlooked in much popular writing. For that reason, this little history stresses *per capita* consumption of energy as well as

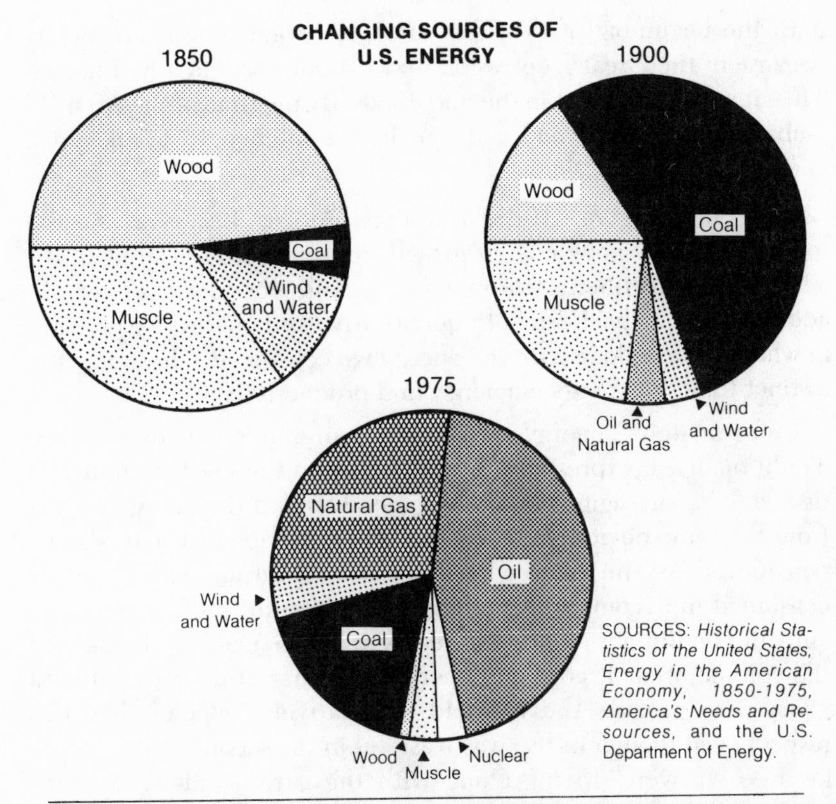

FIG. 1. Sources of U. S. energy have changed drastically during the country's history, as shown in these estimated breakdowns of total energy consumption for three dissimilar periods. Inclusion of some sources often overlooked (e. g., animal and human labor, wind propulsion for sailing ships) dramatizes the transformation from a "developing country" to an affluent society facing enormous new difficulties trying to readjust quickly its traditional balance among legitimate but conflicting goals—affordable energy prices, concentration on readily and reliably available resources, and protection of health, safety and environmental amenities—without seriously disrupting its current standards of comfort and convenience.

per capita output of all goods and services.

The farther back in time we go, the more difficult it is to find trustworthy estimates, but, if we are willing to accept a bit of poetic license before starting the decade-by-decade history, it is possible to draw a fairly graphic overall picture of how energy growth took place. The fuel-consumption equivalents mentioned above can be translated into horsepower-hours. If one then bends the technical definition somewhat to visualize real horses performing actual work on a round-the-clock basis, we can imagine the average energy use of an American during the first 75 years or so of our history as being represented by three or four horses. The overall increase through the second half of the 19th century was less than one might imagine—to something like four and a half horses. From the turn of the century until the beginning of World War II, the total moved erratically because of the Great Depression up to about five and a half horses. Then it jumped to eight in the mid-1950s. By the time of the Arab Oil embargo, each of us had a dozen horses hitched to his personal energy wagon.

This doesn't give a fully rounded picture, however. As the following pages show, energy growth took place by fits and starts. There were frequent traumas—wars and economic crises. In addition, there have been both quantitative and qualitative changes in what we have done with the energy we consumed. This raises the distinct topics of energy efficiency and productivity.

Unfortunately, "energy efficiency" is an ambiguous term today. Traditionally, it expressed the ratio between the energy input to a device (e.g., an engine, motor or generator) and the energy output from the same device. In a broader sense, however, it may also be considered as the degree to which we minimize the energy consumed in accomplishing a given purpose. In the first sense, an auto air conditioner might be rated as to "efficiency" on the basis of the volume of air it cools by a certain number of degrees within a given time period. And it could be relatively "efficient" in this respect even though its use was wasteful in the second sense—either because we were driving along with the car's windows open or because the outside temperature was low enough so that we could be equally comfortable by opening the windows, shutting off the air conditioner, and enjoying the breeze—while using less energy. In terms of the first definition, U.S. history shows a steady record of improvement in energy efficiency—the ability to do things with a given amount of energy. According to the second definition, the various energy-consuming sectors of the nation have had ups and downs over the years.

Productivity is similarly relevant, but also subject to misunderstanding. According to its normal definition of "output per man hour," productivity is not the sole determinant of how big our national economic pie is going to be each year. That is going to depend also on duration of the work week and the size of the work force (which is related to total population, but also affected by other factors—such as age distribution). According to another breakdown, however, gross national product is contingent upon the combined application of capital, labor and other resources, including energy. Thus, energy use is one variable that affects our gross national product, productivity is another, and to some extent these two affect each other. Historically, the increased use of energy has boosted productivity.

As we enter a period in which energy use is likely to face new limitations, it is disturbing to hear from the Joint Economic Committee of Congress that productivity in the U.S. seems to be dropping precipitously on its own at the same time. On an annualized basis, productivity declined during the first two quarters of 1979 by 3.3 and 5.7 percent respectively, with the latter being the sharpest decrease ever recorded since such statistics were first measured systematically in 1947.

In this case, we are unlikely to find specific practical solutions to policy dilemmas by looking back at energy developments in the past, but doing so might help us at least to appreciate the scope of our current and future energy problems. That is why energy history needs to consider such rough reflections of productivity as per capita GNP.

Until now, the major technological trend in U.S. energy has been toward the concentration of resources. Coal concentrates more potential heat in a given weight and volume than wood does, so it could be stored, transported (and—to some extent—*used*) more easily and more efficiently. In quite another way, the construction of dams concentrated energy at a fixed site; but waterpower faced severe limitations until the development of hydroelectricity made the energy product transportable as well. The use of nuclear power reactors and the liquefaction of gaseous fuels have also been moves in the same direction of concentration, and each step of this kind has tended to improve productivity. We have yet to see the effects of a shift in the opposite direction—toward extremely diffuse sources, such as solar energy; and it seems certain that the balancing equation will be complex.

In energy terms we are a poorer country than we were only ten years ago, yet nostalgic musing about our long era of national energy abundance may be somewhat unrealistic. There are plenty of unpleasant recollections from the past, too—boom towns, speculative bubbles, living and working standards that were inferior to today's in dozens of ways.

Nevertheless, each generation assumes (with hindsight) that it could have avoided the pitfalls of the past and magnifies only the advantages. This is one of the dangers of abstract energy planning for the future; and it is one that an evaluation of energy history can help avoid.

In the real world, energy policy depends ultimately on individuals for its full implementation. New technology succeeds in the marketplace only when users are ready for it. The economic, social, psychological and political setting must all be right for inventions and discoveries to penetrate a mass market. Government policies can speed the process in various ways, but policies will be more successful if they take account of natural inclinations than if they are arbitrary.

The remarkable transportation and communication links that now tie our country together may facilitate any new "energy transition," whether it takes place spontaneously or as the result of conscious planning. But there is a counterforce in the sheer number of things that need to be changed in order for transition to occur—the number of residences, vehicles, business units, and so on. An appropriate analogy might be that of a huge, ultramodern cargo vessel; its steering mechanisms are a great deal more sophisticated than those of a motor launch, but its bulk still gives it a far larger turning radius.

Regional and sectoral variations in our energy outlook for supply and demand can be a source of stress in the latest transition we have begun. If such relative differences caused disruptions during a period of low-cost energy, as history shows they did, we might as well expect them again between now and the end of the century. Finally, our relations with the rest of the world—energy exporters and energy importers—will also affect us. International developments have been more important to U.S. energy developments in the past than is generally acknowledged, but the economic and security considerations today seem to be of greater importance than ever before.

What does it all add up to? Where are we headed? By the mid-21st century, U.S. citizens might consider today's misgivings about the fundamental safety of nuclear power ridiculously naive. In fact, they might look back on the use of *non-breeding* nuclear fission reactors as profligate waste—because such systems used only one-sixtieth of the energy power plants are then deriving (safely, of course) from uranium compounds in molten-salt breeder reactors. In the wake of the Three Mile Island accident, such a prospect seems doubtful, yet it really is no more far-fetched than the picture painted by futurists more commonly—of a society based almost exclusively on various types of solar energy. Either scenario is possible. Both are quite possibly wrong.

The straight answer is that even the best general (or individual) cannot anticipate everything in the future. Still, planning is advisable. And *much* of the future is probably right here for us to see—if we try hard enough to interpret the past.

III. 1776-1814: AN INFANT NATION

If American colonists had not won their independence from England a couple of centuries ago, the energy history of this part of the world would have been altered dramatically. Yet the Constitution they adopted was perhaps even more of a history-shaping force than mere separation from the mother country.

In the first flush of independence there existed a relatively sizable minority—thousands, if not tens of thousands—who believed that the fruits of common victory ought to be shared *in common* among all. During the difficult economic period of the mid-1780s their arguments in behalf of what presumably would have become a form of agrarian communism probably had some appeal—as did the noisy mutual challenges of rival states, which brought occasional border fights and even talk of actual war. But the Constitution struck an acceptable balance among various social and economic groups, as well as a workable accommodation between the centralists and decentralists of that day. The federal system that was born during this period—with free enterprise assured and individual protections guaranteed—set the stage for what was to follow. Developments would be regional, but inevitably interrelated. Free-wheeling "Yankee ingenuity" (an historical reality, not a *cliché*) would play a pivotal role in competitively developing and applying the various energy resources available across the continent.

What did we have to work with then?

If England was "built on an island of coal," her upstart colonies from the Canadian border to South Carolina were perched along the edge of what must have seemed like a continental forest. The long-handled felling axe was the first distinctively American tool, and wood was the basic fuel. But wood was not early America's only significant source of energy. Musclepower and—to a lesser degree—windpower also were important.

Energy is defined technically as "the ability to do work," and that can be translated in popular terms into "the capacity to produce change." Burning wood was certainly the most common way of changing *temperature* in those days (to warm a house, cook a meal, help form a piece of metal, and so on). But in that largely engine-less society, fuelwood was irrelevant to the other most fundamental task—that of changing location. Other energy sources had to be utilized in order to move people and things from place to place, to excavate earth, to raise water from wells, to grind materials, and to do all the other jobs that require specialized kinds of motion.

About 92 percent of the population counted in the first U.S. census (1790) was rural, and in any poorly developed agricultural and hunting society a good deal of work must rely mainly on musclepower. Long distance overland travel made use of horses, of course, and draft animals helped with such heavy work as plowing. But farming, commerce and what little manufacturing there was inside or outside the home required direct human labor. Furthermore, in spite of importing African slaves until 1808—when permission to do so via a compromise in the U.S. Constitution expired—the new nation suffered from a manpower shortage. This affected its unstated "energy policy" for generations to come.

For example, Benjamin Franklin had invented his well-known stove during Colonial days but most of his compatriots in labor-scarce America continued to use open fireplaces for heating and cooking. A stove could give about four times as much warmth as a hearth using the same amount of wood; but tree trunks and large branches were virtually free, and the time it took to chop fireplace logs into sticks small enough to feed into a stove could be spent better on other chores. As modern-day energy conservationists are quick to point out, fireplaces are usually inefficient at best, and this was all the more true in view of our forefathers' habit of leaving doors and windows open while stoking the family fire to produce temperatures that left non-American visitors startled and sweltering. Observers from overseas were equally astonished by the technique here during this period of "paving" roads with timber. From the standpoint of long-term resource conservation, using wood for such a purpose may seem absurd now but the method was less labor-intensive than others might have been when the need for maintenance is taken into account. At that time, the same reasoning could have been applied to the "slash-and-burn" mode of clearing land. It was wasteful of wood, but sparing of man-hours.

Another reason stoves were not used more during this period was undoubtedly the fact that they represented "capital goods" in a society where cash for such purchases was generally hard to come by. This relatively primitive state of the economy also makes it difficult for a modern economics researcher to develop precise statistics about the relationship then between energy-input and productive-output. As is the case in the "less developed countries" in today's world, much of the personal real income of early Americans did not go through the marketplace at all. And, because they used relatively little of what we now call "commercial energy," it is difficult to say how much total energy was invested in their home-grown and home-cooked food, their home-sewn (even homespun) clothing, and their family-built housing, furniture and miscellaneous equipment.

With the use of coal and waterpower virtually nil in this country at the time of the Revolution, total energy consumption was almost synonymous with the use of fuelwood and musclepower. Assuming that there were relatively fewer draft animals then but that personal working habits and woodburning practices combined to account for roughly as much individual energy use-up as those in 1850 (a period about which several economic historians have made painstaking estimates), the per capita consumption of energy here in 1776 might have been about one-fourth what it is today.

Comparisons of the "quality of life" across more than two centuries involve too many subjective factors to draw any clear conclusions but, in purely material terms, it seems fair to suggest that average "living standards" have improved far more than fourfold since 1776. To put it another way, we get proportionally more benefits from the energy resources we consume now than people did during the early years of the republic—nostalgia notwithstanding.

Much of what was taking place politically and socially then set the stage for subsequent energy developments. Not the least important was the agreement by the States in 1784—while still joined under the Articles of Confederation—to cede 430,000 square miles to the central government as Federal Territory. The country's doubling of its area in 1803 through the Louisiana Purchase (see Fig. 2) did even more to swell the natural resource bank of the Federal Government. In spite of natural population growth and immigration, the United States found that the number of people per square mile within its boundaries, which largely comprised government-owned land, was

FIG. 2. Much of the United States' historic growth in overall energy consumption has come from rapid increase in total population (largely through immigration) as the nation exploded geographically. At the same time the country remained relatively underpopulated for its area,

practically unchanged more than three decades after we declared independence. We were still an empty country—in which future production, not to mention transportation and communication over the enormous distances involved, would face strict limitations in growth unless musclepower could be supplemented by some other types of energy. It was not at all certain then what they would be, even though Alexander Hamilton noted in his *Official Report on Publick Credit* in 1790 that the country was blessed with coal deposits—which might provide a hedge for the distant future against the day when forest fuel was no longer as plentiful. With such a large percentage of the nation's total territory in federal hands, it is no surprise that a General Land Office (later to become the Bureau of Land Management) was set up in 1812 to administer mineral resources of all kinds from a central vantage point.

We need to remind ourselves that most Americans did not think about coal as a suitable fuel for steam engines during the decades just before and after 1800. In fact many of them did not even think about steam engines at all—except as curiosities. England had not been anxious to share industrial knowhow with other countries (even her own colonies), so technical data about both power tools and the means of driving them were literally smuggled across the Atlantic. Sam Slater's famous mill—America's first—was not built until after the Treaty of Paris had concluded the War of Independence. Even then, it was powered originally by workers on a treadmill, later by a waterwheel.

The first steam engine built entirely within this country was a product of the Revolutionary War and its construction in 1779 might be viewed as a gesture of independence in itself. It burned wood as its fuel, as did Fitch's 1786 steamboat, which propelled itself by mechanical oars rather than a paddlewheel. (Such direct imitation of nature or earlier technology is characteristic of many inventions related to the history of energy. It was probably inevitable later on that our first automobiles would be modeled closely after the buggies they were designed to replace—and be called "horseless carriages.")

Coal *did* have a role in this early period, however—and a critical one! Although its use in stoves and boilers was minimal, it was necessary as a source of coke—which was essential in casting cannon and manufacturing some other munitions. The colonies had imported coal for this purpose from England and Nova Scotia. When war with the Mother Country cut off those supplies, the Revolutionary Army faced a true energy crisis. It was solved by developing domestic mines, including some that tapped deposits George

Washington had noted in his early journals.

Neither of the wars fought with England (one at the beginning and one at the end of this phase of U.S. energy history) was universally popular in this country. Quite apart from the fervor for political freedom, however, there were commercial incentives to seek independence—although these varied from one geographic region or societal sector to another. Frontier farmers at the time of the Revolution looked forward to more effective domestic protection in occasional skirmishes with the Indians. Prospective pioneers in industry wanted to end the various restrictions England had placed on their development. Merchants expected trade opportunities to expand and trading conditions to improve with independence. Although they almost certainly did not think about it in such terms, these and many others who supported the Revolution had ambitions that were likely to be fulfilled only through broader and more intensive applications of energy resources. In this, it may not be too sweeping a generalization to suggest that they differed from many in the sizable minority who opposed the split from Britain as well as those who thought that the War of 1812 was not worth fighting. To a large extent, the "doves" in both cases would have been content with the status quo. The fact that they did not prevail had a broad-ranging impact on our energy history.

Textile manufacturing may have been the first "boom" industry in the new United States. The rapid clustering of factories near the fall-lines of rivers fostered concentrations of population that would maintain their political significance into the second half of the 20th century. Even after steam-driven looms appeared, it was discovered quickly that hungry boilers could produce local deforestation—even in a country that seemed generally to be one enormous thicket. That was an excellent reason for locating steam-powered mills in places similar to those that relied on water power—downstream along navigable rivers. Wood fuel could thus be delivered easily and at fairly low cost. The "mill towns" were born.

The mention of transport costs in connection with wood raises a point that occasional brief references to energy history have commonly omitted. Even though wood in this country might have been free for the taking, it could not serve as fuel until it had been chopped and moved to the place where it would be burned. In the "big" cities of those days—New York, Boston, Philadelphia, Baltimore, and so on—woodsellers charged a variety of prices, but customers could count on paying more for long hauls than for short ones. The chronic shortage of labor was a factor here, too.

It was not accident, of course, that the more densely populated cities mentioned at this point are all ports. Relatively speaking, ships, boats and barges were far more important (both for passenger travel and freight) during earlier periods than they are today. Their movement of goods created jobs; they helped to deliver people to fill the jobs; and then they brought articles of life-support to sustain the people.

Windpower made transocean and coastwise transportation possible. In our times, when energy discussions deal with topics like nuclear fusion and magnetohydrodynamic conversion, it may take some psychological resetting for us to consider sailing vessels seriously as important energy-harnessing devices, but they were. Furthermore, they were not free. There was no fuel cost, but building and fitting ships required considerable capital. Risks were high, but so were the possible profits.

As steamboats began to prove themselves (aided by Robert Fulton's great public relations success with *Clermont* in an 1807 trip up the Hudson), an astute "energy writer" of those times might easily have predicted a bright but limited future for them in the specific field of river transport. There would probably be plenty of wood for their boilers along the banks, even though fairly frequent refueling was likely to be necessary. They could overcome the vagaries of wind and current (*Clermont* had covered 150 miles in 32 hours) and they offered passengers a degree of comfort that surface travelers could never match on corrugated roads. There were still apprehensions about safety but, after all, the hardy inhabitants of the New World were an adventurous lot.

By the end of this period, steam was being studied also as an energy source for land propulsion, and it is interesting that perhaps the most memorable invention of this genre was actually an *amphibian*. In 1805, Oliver Evans (whose name deserves more attention from historians than it generally has received) produced a bizarre vehicle in Philadelphia called *Orukter Amphibolos*—a harbor dredge which could crawl along the shallows of the Schuylkill River, push its stubby floatable frame through the water by means of a paddlewheel astern, or use a system of rocker-arms and pulleys to move it ashore and along Philadelphia streets for short distances on the huge carriage wheels that supported it. Evans chose wood as the fuel for the smokestack-topped boiler aboard *Orukter*, but he foresaw the day when coal might take its place. In fact, he even predicted a useful future for *anthracite*, found in Pennsylvania as early as 1791, but discarded initially as impractical for boilers because it was such a slow-burning fuel.

Evans was one of several inventors during this period whose work provided the basis for truly dramatic progress in the decades that followed. His design of a fully integrated grain mill, for example, employed automation in a way that seems a full century in advance of its time. Its machinery could offload grain from wagons or ships, weigh it, clean it, grind it, and pack the resulting flour into barrels. A waterwheel supplied power to Evans' mill and his intricate arrangement of gears and conveyor belts was a clear invitation to apply steam later on. By combining a variety of operations into a single system, a mill of this type held prospects of overall energy efficiency as well as economy but its eventual success demanded a larger, more concentrated, and perhaps more economical power source than was available then.

Like windpower, the use of energy from moving water in early 19th Century America involved costs that are not always recalled. Apart from the investment associated with the foundations and gearing of the mill itself, operators had to bear the expense of canals that directed river waters through the equipment. The technique employed by a New England textile factory was a good bit more complicated than the one suggested by Currier and Ives prints of individual farm mills, and entrepreneurs who had installed a dam or canal network for their own use would only permit new mill operators to take advantage of it in return for what amounted to a rental fee. As with wind, there was no "fuel cost," but annual "water rights" charges sometimes exceeded the amortised capital cost of building the plant itself.

Nevertheless, the United States economy was moving in a direction that insisted on the substitution of non-human energy (gradually being developed, at a price) for human energy (which was sometimes virtually unavailable in the right form and right place, almost regardless of price). It was the lack of skilled artisans that prompted Eli Whitney in 1798 to propose filling a Federal order for guns by manufacturing them in interchangeable parts. The difference between hand-fashioning each individual weapon and duplicating quantities of nearly identical metal segments which someone else might assemble was the difference between medieaval craft and the modern factory. Once again, however, this may have been a case where an individual inventor was somewhat ahead of the technological and social schedule. Some other manufacturers, who eventually applied power-driven precision tools to the task, became more successful than Whitney in achieving real interchangeability.

All in all, society began to set goals during this period that made it receptive to the technological innovation that accompanied and followed the War of 1812. The country was rich in recognized natural resources, and its citizens were determined to apply those riches at a brisk rate.

IV. 1814-1845: THE AGE OF INNOVATION

Once the second war with England ended, the United States embarked on a period of international peace and internal consolidation. The Cumberland National Road was completed; and canals and railroads bolstered domestic trade through a steadily broadening market. Sensing the trend, President John Quincy Adams tried in 1826 to persuade Congress to finance a major, centrally directed program of internal development; but the move was blocked by his opponents in Congress. Although individual states, sometimes aided by land grants from the federal government, moved ahead in canal and roadbuilding, several aborted attempts by states to create rail systems on a toll-charge basis soon left that field open to private initiative. American production was climbing in spite of the financial panic of 1819. It had started from a very low base, however, and capital was hardly abundant. American canal and railroad projects relied heavily on investment from abroad—including the capital markets of the erstwhile enemy in London. The great building projects of that day relied principally on massive amounts of muscle energy, yet an energy history should also point out that some labor-saving techniques began to appear. For example, explosives were used for earth-moving in the construction of the Erie Canal.

Steamboats came into their own now, and towns along navigable inland waterways flourished. Wood was still the primary steamboat fuel, and many river captains were openly skeptical that coal could ever become a satisfactory substitute. Although the steam locomotive became a more and more common sight, the limited supply of bituminous coal and relatively high coal prices restricted the use of coal-fired boilers before 1840—except on lines adjacent to producing coalfields. Supplies of wood near most rights-of-way were still ample, so there was no need to attempt a fuel switch within a technology that still had other problems to solve. The operating safety of steam engines was sufficiently questionable then that cotton bales were sometimes loaded on a flatcar between the locomotive and the first passenger car to offer the passengers some palpable protection in case of an explosion.

The centuries-old British resentment against coal as a fuel most people considered inevitably sooty and smoky was also a factor in its slow adoption here. Early interest in clean-burning coal systems did not generate much success until considerably later, when the factors of lower availability and higher cost for wood in some areas forced American trainmen to rely more heavily on coal.

"Rail travel" was not synonymous with "steam travel" at the outset. Horse-drawn carriages moved along rails to gain more comfortable and secure routes in a land where road maintenance was haphazard. In cities, the streetcars pulled by horses might or might not be limited to iron tracks; New York had both track-mounted and free-wheeling horsecars as early as 1832.

Many American history books contain descriptions of the famous race in 1830 between an anthracite-burning steam locomotive named *Tom Thumb* and a horsecar moving along a parallel track. The horse won, and, in retrospect, the competition may reflect little more than the fact that contemporary observers do not always focus on the most significant features of a new energy technology. In deciding the economic contest between horse and engine, speed was less important than endurance and the ability to haul prodigious weights and volumes. Much later, in 1860, the Pony Express won back a minor amount of delivery traffic in the West, but the astonishing speed of that glamorous institution enabled it to survive for only about a year and a half before the stringing of telegraph wires along the same route brought about its collapse. While the Pony Express existed, its initial charge of five dollars to carry a half-ounce letter about 2,000 miles in somewhat over a week was cut to one dollar, but even at that rate horsemen would never be serious economic challengers to the locomotive.

Speed may have been a more important factor in regard to sea travel, where windpower was still supreme. The Baltimore Clipper and other vessels of similar design traded hull space for trim lines—an essential asset in outrunning enemy warships or privateers, and a worthwhile commercial characteristic when considering certain cargoes. Perhaps one difference from land travel was that the time to be saved in this case could easily be measured in days, rather than hours. At any rate, wind was barely challenged by steam in ocean transport. The *Savannah* has been hailed as the first steam-powered vessel to cross the Atlantic, and the glory of this accomplishment in 1819 was the reason its name was borrowed some 150 years later for the world's first nuclear-powered commercial ship. But what non-energy-oriented history books rarely add is that *Savannah* used her steam engine for only a very short part of that voyage. All but 85 hours of the 29-day crossing were traversed under sail, and the ship's limited fuel capacity would not have permitted any other course.

Just as wind ruled the waves, wood dominated residential heating, although ever sharper regional price differences began to appear and coal was starting to make some inroads in the commercial and industrial sectors. The direct use of waterpower probably reached its peak about this time, however, providing the driving force for most of over 30,000 power looms that were operating in the U.S by the midpoint in this period. And in 1841 the mighty Niagara Falls themselves were tapped by large waterwheels to provide energy.

Despite the relative underpopulation of the restless new nation, the emergence of Jacksonian Democracy exerted strong social pressure against production systems based on traditionally long work days. President Jackson, whose main support came from the West and South, won added backing among New England factory workers through his 1836 order for a 10-hour day in national shipyards. At this time, Massachusetts factories were still paying only $5 for what approached an 80-hour week, but the tendency toward shortened work weeks without any reduction in pay became a new rallying point for populism. Unless the country was going to stand still (and Americans were in no mood to do *that*), there would have to be some changes.

Coal was by no means an "exotic" fuel to Americans, but the fact is that there was little to recommend it as their first choice for most applications in any sector of the economy. Consequently, its use was spotty. During the War of 1812 a temporary surge in demand for the domestic product had boosted the price of soft coal, and it was only this that prompted serious study of the possibility of using anthracite as an alternative. Roughly three decades were to pass before Americans discovered how rich they were in minable seams of bituminous coal—once people started to look for them in earnest.

Two important uses for coal developed during this period, however. One was in the early blast furnaces, used in the production of iron for some of the steadily growing variety of farm implements and certain industrial goods. The second was in the manufacture of illuminating gas.

One of the cyclical ironies of energy history is that the 1980s may see the birth of a substantial new industry designed to convert coal into a gaseous fuel—even though this is precisely what gas plants were doing in this country well over a century and a half ago. There are differences, of course. The early coal-gas contained far less heat-energy within a given volume than any product suitable to deliver by pipeline today. It was also relatively expensive.

The first natural gas well in the United States was drilled in 1821 near Fredonia, New York, not far from Buffalo. A crude piping system permitted its use in what was then a predominantly rural setting. For the most part, however, natural gas was either unknown, ignored, or feared as a somewhat puzzling fire hazard. Roughly half a century later, when George Westinghouse announced plans to use natural gas as an industrial fuel in Pittsburgh, coal miners opposed it as an unfair competitor, which they believed was likely to rob many people of jobs.

There were some isolated trials of gas lighting in this country by 1812, but the big breakthrough came in 1816 when Rembrandt Peale demonstrated its use dramatically in a museum in Baltimore, near the place the city council met—and quickly won a municipal contract to supply street lights throughout the area. Government leaders and taxpayers must have reacted quite favorably. Indeed, it must have been impressive to see a dark city street illuminated for the first time. Within 20 years, companies had been organized in communities all over the country to produce and distribute gas. At first the service was provided in each area on a flat-rate basis, but practical gas meters were soon devised—making it possible to measure the precise volume of gas delivered to each customer and to bill accordingly.

Although business establishments soon took to using the new gas light fixtures, only the wealthy could afford them in their homes. Furthermore, gas generally was not available outside cities.

In homes, firelight was supplemented by numerous varieties of candles and lamps—with the exact energy source being determined fundamentally on the basis of what was available and affordable. Beeswax made marvelous candles but animal fat and certain types of berries were more accessible to most householders. Whale oil was favored for lamps because it produced a bright, reliable light, but for obvious reasons was not equally available or identically priced in all parts of the U.S. The first synthetic liquid illuminant, camphene, was distilled from vegetable products (turpentine and alcohol). Camphene had the serious drawbacks of a foul odor and a dangerously low flash point. In those days of limited options, however, it became a fairly popular fuel.

Electric energy was still almost exclusively a curiosity during this period—just as steam had been in Revolutionary days. The first patents for the "practical" use of electicity were issued in 1834, but the application then envisioned was to "cure" certain diseases. Although Joseph Henry had constructed a primitive electric motor several years earlier, there was little in his "rocking toy" that seemed to offer hope of becoming an effective substitute for wind, water-power, animal energy or steam engines—much less for candles, lamps and fireplace logs. Yet that did not mean some individuals were not doing their best to promote it. In 1834, Thomas Davenport of Brandon, Vermont, used a battery-powered electric motor to propel a model demonstration vehicle on a small track. The hitch, of course, was the large weight and volume of batteries required in relation to the amount of propulsive power delivered. This became evident on a larger scale when Moses Farmer and Professor Charles G. Page demonstrated passenger-carrying electric locomotives during the 1840s and very early 1850s. As long as the electrical energy source had to be carried along in the form of a heavy chemical storage battery, such vehicles were bound to run out of "fuel" rather quickly. For the same reason, it did not seem likely that electric motors could be kept "fueled" adequately in any industrial operation that would make them worthwhile.

The relative scarcity of musclepower might have been a serious drawback, considering the large land areas available for cultivation, but technology met the demand of the circumstances with a flurry of appropriate inventions. Whitney's cotton gin (1793) had not been a piece of power machinery at all; it was operated by hand. It had already increased the per capita yield of farm laborers on Southern plantations, however, and its economies had probably stiffened the resolve of those in the cotton business to perpetuate the institution of slavery. Cyrus McCormick demonstrated his first reaper in 1831 and patented it three years later. It was followed before long by other horse-drawn equipment that would multiply prodigiously the output of grain farmers—primarily in areas where slavery had never won much of a foothold and where, after 1820, a settler could acquire government land for only $1.25 an acre. The mechanical revolution was essential and effective. Even the early reapers enabled a single man to do the work of five workers with hand implements. Other inventions made similar contributions to other agricultural tasks. A reform of U.S. patent procedures in 1836 gave greater protection to inventors by compelling fuller study to assure originality. This fact probably combined with the emergence of new markets to encourage the flood of innovations.

The new farm equipment continued to use muscle-energy rather than wood or fossil fuel, and in themselves the devices did not require any greater energy input to complete assigned tasks. On the contrary, they raised the productivity *per unit* of human or animal energy input. Nevertheless, the *total* expenditure of muscle energy in this country rose—for the very simple reason that the population was expanding so rapidly. In 1812 there had been between 7 and 8 million Americans (including approximately 1,200,000 black slaves). By 1845—the end of what is called here the Age of Innovation—the total population had risen to more than 20 million and there were nearly 3 million slaves. The number of horses, mules and oxen is less certain, but quite probably it exceeded five million. *America's Needs and Resources*, a comprehensive historical survey produced by J. F. Dewhurst and Associates in 1955, concluded that the work output derived from work animals in the U.S. exceeded that obtained from all types of engines until well after the mid-1800s.

More than 90 percent of this country's slave population was concentrated in the states that were to form the Confederacy, but the issue of involuntary servitude was not the only question that divided the nation along regional lines. After the War of 1812, the infant industries that had been given a boost by the wartime interruption of competing imports pressed for tariffs to protect them from lower priced foreign goods as normal trade was resumed. Manufacturing, however, was concentrated largely in New England and the eastern parts of the Middle Atlantic States. Citizens of agricultural areas saw tariffs as unnecessary additions to the cost of living, at best. At worst, Southerners and Westerners saw tariff barriers as an invitation to reprisals by other nations that would damage U.S. exports of cotton, tobacco, and other non-manufactured products. By 1833, U.S. agricultural interests, representing a way of life that was less energy-intensive than manufacturing, forced some significant tariff cuts, but the battle was not finished. It had barely begun.

Land in the West continued to beckon to settlers, offering natural energy resources in the form of wood, wind (for mills), and running water that were more than ample for simple and isolated homesteading operations. The development of communities came more slowly, however, and lamp-fuel or anything else that came from the Atlantic Coast commanded a dear price. Land speculators moved west too, and get-rich-quick schemes sometimes fell apart in a violent way that shook the whole country—as in the financial crisis of 1837. Nevertheless, the spread of population toward the Pacific was inexorable—especially after the publication of Fremont's reports on the vast country of the Rockies. Although his renown as "the Pathfinder" could not win him the Presidency when the newly formed Republican Party nominated him as its first candidate in 1856, Fremont was an authentic national hero. His explorations of the Far West in 1842 would be even more of a blueprint for the future than those of Lewis and Clark had been earlier—especially by replacing old fears of a Great American Desert with accounts of fertile and plentiful farmland.

What may have surprised Fremont was the way in which the new push westward came. In the long run, a good bit of what happened thereafter was a result of a single electrical invention that appeared near the very end of the Age of Innovation—Samuel F. B. Morse's telegraph. Although it was similar to other electrical devices developed up to this time—in that it was tied to bulky wet-cell batteries as a power source—that was no great handicap for the telegraph. It operated from a fixed location where a crew could service its "fuel" demands of battery plates and acid, although at a not inconsiderable cost. The telegraph did not heat anything; it did not propel a vehicle; and it did not energize equipment. Yet the rapid application of this invention was an instrument of social and political change for the entire nation in the decades that followed Morse's installation of the first full-scale telegraph line between Baltimore and Washington in 1844.

V. 1845-1880: TURMOIL AND TRANSITION

The years 1845-1880 were some of the most volatile and violent in all U.S. history. Greatly enhanced communication exaggerated changes. News flashed from one place to another like lightning. Broad public response to events developed with unprecedented speed.

The telegraph, widely circulated print media, and improved photography were to this period what television would become a century later. The annexation of Texas and war with Mexico swiftly pushed the American frontier westward again—by virtually half a continent. Midway in the period, the country was split by a civil war that shook the national economy to its foundations and—far more tragic—left the bloody imprint of about a million casualties. Some 500,000 fighting men died of wounds or disease, and perhaps an equal number were either disabled or uprooted permanently and reduced to wandering aimlessly for the rest of their lives.

From a political point of view, matters were just as jumbled. The first U.S. President to be assassinated was succeeded by the first to be impeached, and he was followed by an administration in which blatant corruption invited more public disillusionment and unrest. With all of this, it is easy to see why the revolution in American energy use that took place during these same 35 years drew relatively little attention from historians.

Astonishingly, the average rate of population growth during this period of turmoil and transition remained almost as high as during earlier years. If this had been due entirely to the increase of births over deaths, the rate of increase would have been comparable to the one in Mexico today—which gives so many population experts such concern. In the case of the U.S., however, there was a great surge in immigration, beginning in the late 1840s. Furthermore, the simultaneous growth in national territory kept internal population pressures from developing; actually, the number of inhabitants per square mile *declined* between 1840 and 1850 (See Fig. 3). By 1880, despite all the upsets the nation had suffered, the potential clearly existed for the largest marketplace the world had ever seen—unmatched anywhere in its combination of united national area, diversity and richness of resources, communication and transport facilities, and total number of ordinary individual citizens with significant spendable income.

Within the narrow perspective of energy history alone, railroads undoubtedly were pivotal. They triggered what is usually labeled "the nation's switch from wood to coal."

So long as rail travel was limited to relatively short runs in the

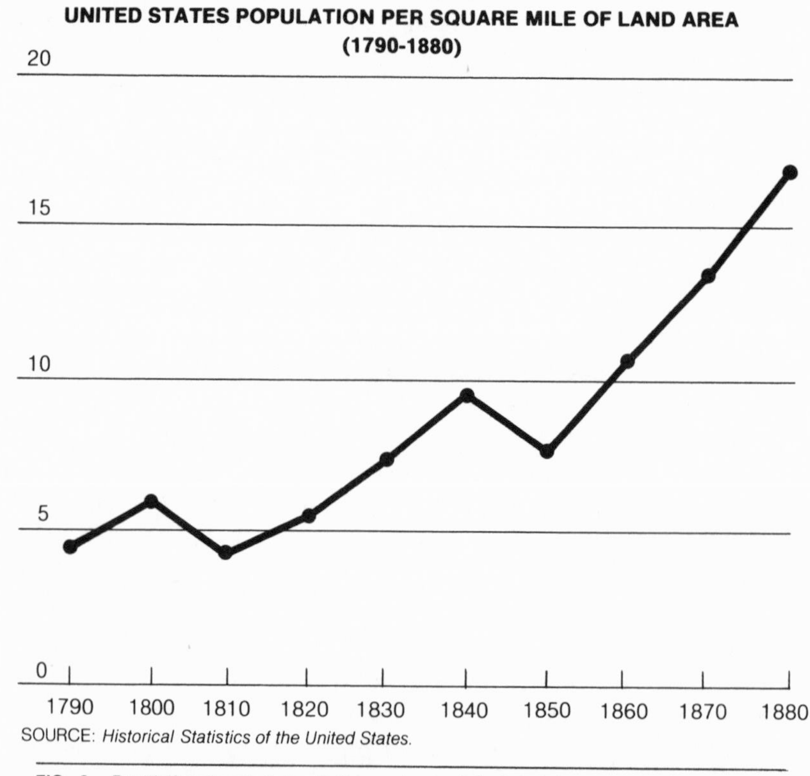

FIG. 3. Population density in the U. S. remained relatively low throughout the century or so it was a "developing country," and the statistic even dipped on a couple of occasions as the nation acquired new territory. By the end of the First World War, however, the 1880 density had doubled. Despite the Great Depression and the introduction of birth control as U. S. mores changed, population density doubled again by the late 1960's. Today, after the admission of Alaska and Hawaii as states, it still stands at only slightly more than 60 persons per square mile—about one-tenth that of such countries as Japan, West Germany and the United Kingdom.

general vicinity of the Atlantic coast, wood was the obvious fuel for most locomotive boilers. Once the nation made a commitment to press into areas where forested land was less ubiquitous and where cheap water transport was unavailable in the directions people wanted to travel, however, the choices between wood and coal had to be reevaluated. Coal won on several counts.

When good quality coal is burned it produces about four times as much heat as the same weight of top-notch fuelwood. If equal volumes are compared, coal has an eight-to-one energy advantage. Thus, to begin with, coal greatly extended the range (and/or payload capacity) of locomotives designed to haul their own fuel on long trips. In addition, it was soon discovered that coal deposits existed near some of the new railroad rights-of-way where wood supplies could have become a serious problem.

Once the two fuels were regarded as more or less interchangeable, price became the only factor that enabled wood to hold on as a locomotive fuel where it did. Still, regional diversity was far greater than many modern writers assume, as an excerpt from a *Handbook of Railroad Construction* published in Boston in 1857 makes unmistakable:

> "It does not follow that because coke in England, anthracite in Pennsylvania, or wood in New England, is the most economical fuel that either of the above will be so in Ohio, Indiana, or Illinois, or because wood is the cheapest in some parts of a state, that it is so throughout, or even that one fuel should be applied to the whole length of a single road."

If train boilers had been the only new market for coal at this time, however, the quest for new mines and more efficient exploitation of existing ones probably would have gone on at a slower pace. The average price per ton might have remained higher, and it would have taken longer for coal to achieve its full competitive economic advantage over wood. Other technological developments during the same period, however, pushed the coal industry to even greater growth.

Building a national railroad network involved the laying of thousands of miles of track. Metallurgical standards for rail were not as high as those for the metal tools which previously had represented a large share of the demand for iron, and the need for rails on a larger scale occurred at almost the same time new manufacturing processes managed to improve the quality of iron produced with coke (made from coal) instead of charcoal (made from wood).

Because coke was more economical—and likely to stay so because trains could deliver enough of it to produce a given amount of heat energy more cheaply than the bulkier charcoal—old and new iron works were more than willing to make the shift. The introduction of steelmaking to the U.S. in the 1850s and 1860s confirmed the industrial fuel change; but the continuing demand for rails (fabricated by that time of steel rather than wrought iron) dictated the manner in which pig iron was to be produced and the Bessemer steel process was to be applied during the 1870s. A sidelight worth pointing out is that the lower relative cost of coke was not always due to a scarcity or high price for wood; charcoal was expensive primarily because its production was labor intensive.

Certain other spin-offs from rail expansion also benefitted coal. Throughout industry, cheaper mass production of metal parts replaced older boilers—which originally had often been made almost entirely of wood, except for the fireboxes. Experience and experimentation with steam engines for trains helped to spawn safer, more powerful, and more efficient stationary engines. Those were the logical power sources for industry and commerce in the new population centers that blossomed around railheads as the iron horse rolled toward the Pacific. Barely 20 years after Californians raised the American flag for the first time, their state was linked to the original colonies by a transcontinental rail line which nourished the intervening string of tree-poor towns with coal.

Expansion of the rail system was popularly endorsed as a national mission. When trains between Boston and Fitchburg, Massachusetts came within 500 yards of Walden Pond, even a nature lover like Henry David Thoreau was moved to find cheer in "the muffled tone of the engine bell . . . . which announced that the cars *are coming* . . . notwithstanding the veto of a New England northeast snowstorm." It may even shock some of Thoreau's latter-day admirers to learn that he enjoyed the fact that trains' comings and goings had become "the epochs of the village day" and endorsed the way "one well-conducted institution regulates a whole country."

The great interstate railroad systems were not built by government, but they surely would not have developed so rapidly without considerable federal assistance. The biggest boost came in outright grants of land for the rights-of-way—over 130 million acres valued then at about two billion dollars—but there were also indirect subsidies in the form of troop protection and lucrative government shipping contracts. As for risk capital, a great deal still came from overseas. Adverse financial developments domestically, such as the panic of 1873, could turn it off like a faucet.

Nor did the U.S. government enter directly into fuel production in any major way either—although Congressional legislation began to affect it, usually in a sympathetic manner. The Mining Law of 1872 spelled out ground rules for the use of all public land resources and the U.S. Geological Survey (founded in 1879) contributed important reference data through its continuing studies of the national mineral potential.

The great increase in America's use of coal did not result primarily from its being substituted generally for wood in all applications that were already underway. More fuelwood was burned in the United States in 1880 than in 1845—at the very time the nation was "shifting from wood to coal." The fact that the rate at which consumption of wood increased did not match the burgeoning population growth might easily be attributed to the fact that the urban population of the East was finally starting to shift significantly from fireplaces to stoves for heating and cooking. Instead of felling trees in the wilderness and dragging logs home as their forefathers might have done, city dwellers of the third quarter of the 19th century bought precut fuelwood—so the convenience of stoves was more appealing. The influence of comparative resource abundance and labor scarcity could still be noticed, though. To save time and reduce costly manhours, Americans used broad-bladed saws that Europeans considered wasteful.

The primary energy transition that took place during this period was in the consumption *sectors*. From the beginning to the end, more than nine-tenths of all firewood was used in households, including—especially in the earlier years—such domestic industry as the curing of meats and tobacco. But what we now designate as the "industrial" sector began to grow at a more rapid pace than household use, and coal fueled much of that new expansion. Meanwhile, the transport sector *really* mushroomed. The central factor was the rail system. From developing *half* as much power as the nation's factories around the middle of the century, steam locomotives went to using more than *twice* as much as factories by 1880.

Regional differences must always be recalled, and that brings us back to some of the socio-economic-political tie-ins. The Pocahontas coal fields of Virginia were not exploited until after the Civil War, so the Confederacy had to use wood in its train engines almost exclusively—even though it was hurt by the consequent loss in carrying capacity. The longer growing season in the South helped replenishable wood fuel to hang on longer than in other areas; wood-burning locomotives continued to be used regularly in Florida until around 1900. But the regional scarcity of coal had serious repercussions between 1861 and 1865 when Southern states tried to match extra wartime demands with suddenly reduced supplies. Railroad beds themselves had to be cannibalized so ties could be used to fuel boilers, and cottonseed was substituted for coal in the production of illuminating gas. Northern forces, on the other hand, had reasonably secure supplies of coal. Nevertheless, wartime inflation pushed prices up everywhere, and in 1863 the New York Central reverted to wood temporarily for purely economic reasons.

From a political point of view, the nation's sudden surge to the West had probably helped to precipitate the Civil War. Although the California Gold Rush was initially an instigating factor, a more sustained incentive was the lure of large areas which could be turned into independent farms and ranches. With Fremont as their first Presidential candidate, the Republican Party eventually coalesced from groups that favored: a) free land in the West for homesteaders, b) federal encouragement of the communication and transport links that would support them, and c) the abolition of slavery, which was repugnant ideologically and worthless to them as an economic institution.

After the War Between the States had ended, the relative shortage of workers persisted. The heavy labor force for the railroads was recruited in considerable measure from among both European and Oriental immigrants, and agriculture turned to another source of energy. Reel-and-cable systems were developed so stationary steam engines—fueled either by coal or wood, depending on local conditions—could draw farm equipment back and forth across fields. In time, steam-powered harvesting equipment was mounted on wagons drawn by horses. The emphasis in new inventions was on minimizing the amount of human muscle-power that would be required. Farmers still might have backbreaking jobs, but their individual productivity increased further—albeit at the expense of raising the capital investment a farm owner might require to get started. Even the introduction of a relatively simple new product like barbed wire (1874) had a powerful long-range effect on energy consumption. Barbed wire spelled the end of open grazing. It protected homesteaders against the ravages of herds that had been allowed—even encouraged—to roam freely. That signaled the broad triumph of farming (which was growing steadily more energy intensive) over open-range cattle ranching (which occupied much land with relatively *small* numbers of people and *little* total expenditure of energy).

Here, however, a point made earlier needs to be given added stress: At mid-century, primarily because of our relative emphasis on agriculture, the amount of human and animal energy that went into plowing, digging, hauling, etc., was still about five times as great (in horsepower hours equivalent) as what *both* wood and coal were contributing to the smaller industrial and transport sectors of the U.S. economy. By 1880, though, the manufacturing-and-transport sector, with an increasing reliance on coal, had replaced farming in terms of the nation's prime interest and commitment. That change in emphasis for consumption sectors was the real energy transition we went through.

Throughout the 1845-1880 period, waterpower maintained an important position in sections where it was available. Shortly before 1850, water turbines began to replace the old overshot and breastwater wheels, increasing efficiency. In 1849, a wooden dam more than 1000 feet long and about 30 feet high was built at South Hadley Falls, Massachusetts. Sophisticated diversionary canals were being installed in various parts of New England around the same time to extend the natural power source beyond the original limits of the riverbanks. Once such large capital investments had been made in waterpower, it would have been foolhardy to abandon it in favor of steam engines that would require newly trained personnel and might cost twice as much to operate. Thus, it was only in instances where *additional* power capacity was needed that steam was even considered for the area, although one of the advantages a cost-conscious Yankee entrepreneur might note was using the engines' "waste heat." In the cold New England climate, an operating engine helped to keep the entire factory warm.

In the South, a pattern of sharecropping that replaced slavery kept most areas from diversifying enough to become self-sufficient, but in some cases Reconstruction was accompanied by industrialization. Coal and iron deposits discovered in northern Alabama and Tennessee helped Birmingham grow out of nowhere within a couple of decades. Low prevailing wages and an abundance of waterpower encouraged textile mills in Georgia and the Carolinas to expand during the 1870s—offering serious competition to New England for the first time.

Waterpower's major failing in this pre-electric era, however, was its inherent immobility. The country had simply spread out too much for even the *improved* water technology in a few areas to keep pace with national growth and accompanying new energy demands. By 1868, steam equaled waterwheels as an overall power source for U.S. industry.

Windpower gave a great final gasp before it also faded as a major factor during the last quarter of the 19th Century. Windmills may have seen only limited use in the colonies and the more compact communities of the early Federal period, but they made sense on large farms of the type fostered by the move westward. Elaborately geared systems could pump water, thresh grain, and mill flour. And at sea, America's sailing vessels still outnumbered commercial steamships at the time the United States marked its Centennial. Wind-propelled vessels had matured in size to thousands of tons and in speed to peaks of around 24 knots. There were still many who believed steam could never be more than a supplement to sail for trans-Atlantic travel.

The two developments that ultimately tipped the economic balance in favor of steamships seem at first glance to have little to do at all with energy policy. One was the introduction of metal hulls, which made very large ships possible and thus increased the reasonable capacity for fuel as well as the overall space for cargo. The second was popularization of the screw propeller, which converted steam power into motion through the water far more efficiently than paddlewheels. The greater efficiency of the new ship propellers meant less steam (and thus less fuel) would be needed to make an ocean crossing—even at constant speeds comparable to the overall trip averages for sailing ships.

Because of its greater energy density, coal faced little competition from wood during this stage as the choice for long-distance ship fuel. During the Civil War, petroleum had also been tried with considerable success in ships' boilers and it promised even greater compactness and potential handling convenience. Analytically, it offered improvements in both speed and endurance, but in 1867 Navy Secretary Gideon Welles rejected the whole idea of using petroleum on steam-powered vessels. He argued against it on the grounds of health and safety, criticized it as *inconvenient* since oil then was transported mostly in barrels and wooden tanks, and suggested the price of petroleum made it impractical.

We must recall that when Colonel Drake had drilled his historic well at Titusville he had not been looking for an engine fuel. He was seeking a substitute *illuminant*. The many lamp fuels available then were generally expensive; and the more affordable ones had the additional drawbacks of being dim, smoky, smelly or explosive (or perhaps some combination of those). Kerosene, a trade name applied popularly to a variety of liquid fuels, including some produced from coal, looked like a possible winner and was given a fair amount of commercial ballyhoo. Total victory in the market place, however, depended on uncovering a dependable source of some crude liquid base that could be refined more or less directly into an acceptable form of kerosene lamp fuel at reasonable cost. Before 1859, petroleum had been a byproduct of salt wells and some natural springs. The significance of Drake's Pennsylvania venture was that it was the first commercial effort that produced petroleum *from a well of its own*. Others followed suit, and an oil rush was on.

Although the production costs of the earlier lamp oils had varied greatly, the retail prices were often close together. In a broad sense, potential demand exceeded the combined supply; and the different products tended to be nearly interchangeable where they were jointly available. The rail network was in its early adolescence and road travel was still arduous, so transportation costs were a major factor. In such a setting, it is understandable that the price of crude oil at the wellhead went through wild gyrations over the first few years. In 1860 it was nearly $20 a barrel—about the same as the ingredients of camphene on a per-gallon basis but much lower than sperm oil and lard oil. As more successful petroleum wells were drilled and difficulties developed in delivering the product to market, the wellhead price dropped all the way to ten *cents* a barrel during 1862. By the end of 1864 it was back to $11, and improvements in transporting crude quickly helped expand the market. By 1865 trains were pulling wooden tank cars filled with oil—to be followed in four years by better ones made of iron. These were the days of cut-throat competition and by 1875 a major rate war was in progress as one rail line or another tried to obtain a monopoly in the transport of oil. By that time, seagoing oil-tankers had evolved from barrel-carriers to vessels with specially constructed hulls and the U.S. had become a petroleum supplier to the world. The expression, "oil for the lamps of China," traced its origins to America's wells. The first large pipeline did not cross the Appalachians until 1878, however, and that was in the era when oil moved in this country primarily from east to west.

Like electricity, petroleum had been touted in the earliest days for "medicinal" properties. The hustling of "snake oil" salesmen gave rise to still-familiar expressions and mental images. By the 1870s, however, petroleum-based pharmaceuticals such as vaseline, lotions, etc., were showing genuine value. The use of "rock oil" in lubricants for machinery gradually developed too, but the major application was definitely in bringing more light to households.

Gas had gone through various phases as an illuminant in this country. A patent to distill burnable gas from refuse had been issued in the U.S. as early as 1815; and although coal was the primary base throughout the 19th Century, the fuel also was manufactured commercially in various localities from many different products (including pitch, pine-oil, petroleum, and—before it became outrageously expensive—whale oil). Many of the process modifications were based on what would now be called "environmental concerns," such as interest in eliminating odors. Some were aimed at reducing costs—in response, for example, to an intensive popular campaign in New York during the early 1850s to reduce gas rates so "every private home" could have cheap illumination without resorting to the dangers of a fuel like camphene. Nevertheless, even at $3.00 per thousand cubic feet, a typical "target rate" of the late 1850s, gas did not seem capable of competing with lamps using kerosene refined from petroleum.

Much of the progress in energy technology the United States made during this tumultuous period was evidenced in the gigantic birthday party the country held for itself at Philadelphia in 1876. Trains served the Centennial Exposition site. Total attendance was equivalent to about one-fifth of the country's population. In the largest U.S. pavilion, a 6,000-horsepower steam engine drove acres of machinery. There were kitchen ranges operating on lighting gas, equipment for farming, and engines fueled by a petroleum "waste product," called "gasoline." Alexander Graham Bell showed up one weekend to demonstrate his battery-powered telephone. Not even the latest expansion of human energy was neglected: Female Americans had come a long way since the first "women's rights convention" nearly three decades earlier and had their own pavilion at Philadelphia—featuring a woman engineer.

As might be expected, gas lights were an integral part of the exposition. More of a novelty were the brilliant electric arc lights. These huge fixtures were simple in principle but cumbersome. A high-voltage spark flashed between two electrodes and a reflector focused and reinforced the light, turning the darkness of night into an eerie approximation of daylight on the midway below. The carbon rods that served as electrodes were consumed steadily by the heat, so they had to be advanced on a precise time schedule to keep the gap constant. After a relatively short time, the long graphite rods had to be replaced.

An arc lighting system clearly was suitable only for special purposes, such as street illumination, and did not seem competitive with either gas or oil lamps for home use. Its power demands were huge compared with today's systems. Nevertheless, arc lights were an eye-bulging example of what the centennial motto promised the U.S. and the world—"the best that we can do." Three years later, in 1879, Thomas Alva Edison produced a practical electric lamp using a different principle. That was one of a series of related inventions that helped transform the country over the next few decades into an economic titan.

VI. 1880-1918: NO LONGER A "LESS DEVELOPED COUNTRY"

Having survived its boisterous beginnings, the nation exploded industrially in the next three or four decades. It emerged as a cocky giant, with coal smudges on its face.

It is not the purpose here to assign specific shares of blame or praise to government, business or labor for the growing pains the United States suffered in the process. What seems indisputable is that by the time this transitional period ended *big* government, *big* business and *big* labor were all firmly established. For better or worse, energy developments in the U.S. from this point forward were likely to be affected as much by these three super-forces as by the still necessary summation of many, many distinct individual choices.

Two financial panics interrupted the era—in 1893 and 1907. One school of thought is that both were caused essentially by reckless industrial overexpansion, but other writers lay the fault in each case at the doorstep of the federal government—for allowing gold reserves to be drawn below the legal minimum to placate "soft money" agricultural interests in the South and West, and for pressing Teddy Roosevelt's trust-busting policies and weakening the whole economy. In the final analysis, whether the fault was industry's or government's, the nation managed to pull itself together only by a truce between the two. In 1893, in fact, the Morgan banking interests literally refinanced the national government.

In a nation where a population of employees was speedily superseding the old pattern of small, self-sufficient family farm units, some sort of labor organization was inevitable. The labor union movement may have been strengthened by the realization that economic swings could be felt swiftly and forcefully—as well as by the government's demonstration that it would intervene in the marketplace if there was a public clamor for reform. Work stoppages had occurred before in protest over working conditions or wages, but during this period strikes were numbered for the first time in the *tens of thousands.*

All of these developments affected the way energy was produced and applied.

Shorter working hours, better pay, improvements in working conditions, and the enforcement of new quality standards to protect public health and safety all tended to add to basic production costs—motivating a general quest for counterbalancing improvements in production methods. Fresh concern about resource limitations and protection of the natural environment—usually associated with the first Roosevelt, but actually traceable to literature of the late 19th Century—also prompted new targets for energy efficiency.

For the most part, steady technological advances supplied the key to success. It was also during this period that the dramatic tinkering of individual inventors began to give way to more specialized research "teams." The results included many products—commercially affordable aluminum, hydrocarbon-based synthetic materials, and gasoline additives, to mention but a few. The relationship of these products to energy consumption patterns would not be felt until later, however, when U.S. involvement in two global wars would modify American living habits still further.

As for government's role, the perceptible trend during what has been called the "Progressive Era" was toward direct regulation. In the last part of the 19th Century and the beginning of the 20th, rail and oil monopolies were attacked and eventually split apart. New taxes were levied, and the supervision of interstate commerce developed from a fairly vague constitutional theory into a day-to-day reality with establishment of the Interstate Commerce Commission. The biggest step of all came as an emergency measure as this country became involved in World War I: The national rail and communications system was taken over temporarily by Washington, which also assumed direct control of the production and distribution of fuel.

In the midst of all these separate (and sometimes conflicting) developments, the underlying trend between 1880 and 1918 was that the United States economy itself grew larger and stronger and more complex.

Production of pig iron was a fundamental indicator of industrial growth, because it is a basic ingredient in so many manufactured products. At the start of this period, the U.S. was turning out less

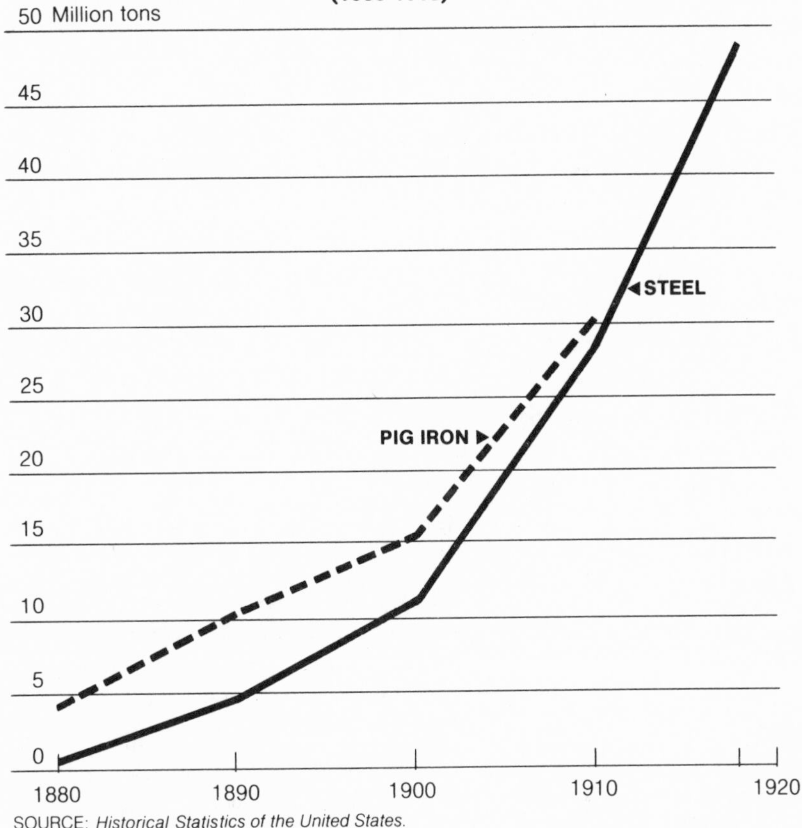

FIG. 4. The rapid transformation of the United States from a "developing country" to a world leader in industry was reflected in its output of basic metals. For the sake of comparison, the United Kingdom started this same period with an annual production of 8.68 million tons of pig iron and 1.46 million tons of steel; but by 1918 that country was turning out only about one-third as much of the first product and one-fourth as much of the second as the U. S.

than half as much pig iron as the United Kingdom. In less than ten years the U.S. was the world's leader in that field. The nation zoomed to first place in steel too. Output in that case increased even more dramatically, multiplying more than 40 times between 1880 and the end of the First World War (See Fig. 4). By 1900, the United States had welded much of the steel into an ever more extensive and intricate rail network, whose 200,000 miles matched the transport facilities of Europe.

Growth fed on energy and by 1918 over three-quarters of the U.S. energy diet was coal. During the same time, the relative importance of wind and animal energy faded to insignificance. In 1918 the number of horses and mules in this country peaked at 26 million. This seems like a very large number, but by that time the nation had reached the point of measuring overall work output (mostly from engines and motors) in *trillions* of horsepower hours. Meanwhile, waterpower was born again in the form of hydroelectricity and by the end of this period hydro was responsible for about one-third of all U.S. electricity. Even so, it still accounted for only a few percent of the country's *total* energy use.

It was primarily blast furnaces and boilers, stoked overwhelmingly with coal, that ate up energy as never before. Writers who dramatize America's increasing appetite for energy within the 1950s, 60s and 70s should look back to the decades bracketing 1900 if they are really fascinated by growth in total energy consumption. Population more than doubled between 1880 and 1918, but per capita energy use also doubled. The overall effect was a rough *quadrupling* of energy use.

The underlying reason for the growth in consumption was not that American society suddenly became more energy-intensive; the stage-setting changes in economic sectoral emphasis had already largely taken place. Instead, it was because the nation began to *produce* more—a lot more! The statistical evidence here is easy to misread, however, because of the tendency to forget how important wood had been as an energy source before coal succeeded it.

To a very great extent, the direct substitution of coal for wood finally took place across the nation during these decades, even in residential and commercial use. If the heat content of the wood used annually by Americans is factored in (it dropped from nearly three cords per person in 1880 to substantially less than one cord by 1918), it suddenly becomes evident that *total* energy consumption (coal, oil, gas, hydro *and* wood) was closely tied to the rise in real gross national product during the period. Omitting wood from the calculation (as recent articles about energy usually do) gives a false impression of the degree to which Americans increased their individual energy appetites at that time.

It is difficult in fairness to denounce this country's actual energy pattern at that particular stage in national development as wasteful—a point we might keep in mind when considering the energy aspirations of developing countries today. If real GNP within the U.S. increased about fourfold at the same time the population merely doubled, energy sources appear to have done an important job with reasonable efficiency. Largely as a result of higher energy use per capita, twice as much material product became available potentially to each American.

This is not to say that growth took place without inequities and some largely unperceived trade-offs. For one thing, essentially unregulated coal burning and the increasing concentration of population in urban areas combined to produce serious pollution problems. That did not seem to inhibit either community or national pride, however. Many a letterhead and government emblem during the period featured smoke billowing from the stacks of factories or locomotives. Regardless of how we have come to regard such matters within the past generation or two, Americans then flaunted all the badges of their emergence as a "developed country."

Internationally, the U.S. rather suddenly became a force to reckon with. To the enthusiastic applause of a sensationalistic and extremely nationalistic daily press, the nation showed off its new power in the Spanish-American War and by brandishing the Great White Fleet around the world. It was not that the country was purely aggressive; in fact, Teddy Roosevelt played the role of international peacemaker to help settle the Russo-Japanese conflict. Almost naively, the United States of America merely sensed its new might and wanted to be sure others did not overlook it.

There is appropriate symbolism in the fact that the U.S. Centennial Exposition took place in the old Colonial port city of Philadelphia, but that the next grand event of this type moved to mid-America. The atmosphere at the Columbian Exposition of 1893 in Chicago was different—prouder and more outspoken, yet advertising accomplishments of the *present* as much as courage of the past or promise of the future.

From an energy standpoint, the central attraction this time undoubtedly was electricity. President Grover Cleveland opened the exposition from Washington by pushing a "magic button" which sent an electrical impulse by wire to set in motion a huge Allis engine in Chicago and deliver power to the key exhibits. There was an "Electrical Building," in which a "majestic luminous column" was studded with thousands of electric incandescent lamps.

Three more major international expositions were held in the U.S. between the turn of the century and the Great War. Each showed off new extensions of technology and new energy applications that were then taking place at a breathtaking pace.

The Pan American Exposition of 1901 in Buffalo took advantage of the hydroelectric installation at Niagara Falls, which had begun operating in 1895. The dominating feature was a 375-foot electric tower, decorated with incandescent bulbs and searchlights, and topped by a statue of the "Goddess of Light."

In 1904 the fair at St. Louis to commemorate the centennial of the Louisiana Purchase highlighted the applications of "science of everyday life," including new kitchen ranges. Its crowning feature was a display of 100 automobiles—the first sizable exhibition of this type.

The Panama-Pacific International Exposition of 1915 was a salute to the opening of the Panama Canal, and it was supposed to focus on cultural contributions. Yet, a great many visitors remembered it primarily because it was the first to offer airplane rides to the public.

A cursory review of such milestones in energy might suggest that new technology replaced the old overnight across America. It did not. It is true, for example, that electric lighting quickly became available in all large cities and most small ones, but gas gave it vigorous competition throughout this period. The popular introduction of the gas mantle in the late 1880's offered six times as much illumination from the same fixture while actually reducing gas consumption, so there was no clear economic incentive for home owners to convert to electricity. Furthermore, gas mantles gave whiter light and less glare than early incandescent bulbs.

As happens often with new technological developments, electric current also aroused public fears about safety. The apprehensions were undoubtedly reinforced by Thomas Edison's own fierce personal attack on George Westinghouse's ultimately successful efforts to switch from direct current (DC) to alternating current (AC) as an easier and more economical means of transmitting power over long distances. Edison went so far as to publish an article in 1889 on "The Dangers of Electric Lighting," wherein he characterized AC systems as unreliable, hazardous to life and property, and aimed only at trimming electric company outlays for "copper wire and real estate." He is also quoted as having said that "Just as certain as death, Westinghouse will kill a customer within six months after he puts in a system of any size."

Apparently, such warnings troubled President Benjamin Harrison. Harrison has been hailed as a pioneer, because he had the gas chandeliers in the White House converted in 1892 so that they could use either gas light or electricity, but secretly he was afraid of being shocked if he touched the electric switches. The President issued a standing order for an aide to turn the downstairs lights on at dusk each evening, and they stayed on until that staffer returned the next day. In their private quarters, members of the first family discreetly continued to use gas.

Electric power got an extra boost in this country because a host of useful gadgets apart from the light bulb linked it simultaneously to other conveniences. The electric flatiron and electric fan were invented in 1882, to be followed by various types of electric ranges in the 1890s and the electric vacuum cleaner and washing machine in 1907. The automatic toaster popped up in 1918.

Nevertheless, it took quite a while for electric motors to replace steam-powered equipment or other machinery in industries that burned fuel directly on site. In the 1890s Benjamin Butterworth, a U.S. Patent Commissioner who chronicled industrial growth, wrote that the electric light had already been fairly well established for some time, but electric motors had "not yet been sufficiently demonstrated to be cheaper than coal."

In a sense, of course, most electric motors were actually *using* coal, and lots of it—albeit indirectly. Edison's Pearl Street Station in Manhattan had to burn 10 pounds of coal at first to generate each kilowatt-hour of electricity. Even by the first decade of the 20th Century, the most efficient generating stations were still using about six pounds per KWH. (Today, it takes about a pound of coal to generate a KWH.)

The ultimate advantage of electric power in either residential or industrial applications would be appreciated gradually in terms of its transportability—and thus flexibility. Wires could reach almost anyplace, so fewer and fewer people had to be isolated from sources of energy for nearly any application—heat, light, or mechanical motion. Factories would be able to reorganize production in any layout or sequence that contributed to efficiency—without being limited to straight-line systems of belting and drive-shafts that transmitted mechanical power from centrally located prime movers. The most dramatic outcome of this new approach to manufacturing was epitomized by Henry Ford's technique of moving his production line of cars in a steady flow past workers in fixed locations.

At the turn of the century a futurist might have sensed that the automobile was a potentially significant invention, but few guessed its ownership would become so widespread and its use so central to modern life. After all, the capital investment involved was fairly high and alternate forms of transportation were abundant, even for a society that had as much relative individual affluence as this one. For cross-country travel, rails existed and decent roads generally did not. *Harper's Weekly* commented flatly (August 2, 1902) that "The actual building of roads devoted to motor cars is not for the near future, in spite of many rumors to that effect."

In cities, cars seemed unnecessary because the electric streetcar had appeared during the 1880s. By 1905 New York had even opened its first subway, powered by a "huge"—for its time—steam-electric generator.

It was not clear then which basic *type* of automobile might eventually dominate whatever national market developed. Slightly over half of the 8,000 automobiles registered in this country in 1900 had been built here, and most were steam-driven. Battery-powered electrics were in second place, with the internal combustion engine bringing up the rear. What may have decisively tilted the odds in its favor was the opening of the enormous Spindletop Oil Field in Texas in 1901. Just as the first oil drilling in Pennsylvania had provided ample new lamp fuel, so this new discovery seemed to promise plenty of cheap petroleum for as many motorcars as America wanted. It turned out, however, that America wanted quite a few—a million by 1913 (after the Model-T had begun the mass production that reduced prices) and five times that many over the following five years.

Automobiles fit right into the independent, free-wheeling spirit of Americans. Motorcars were highly personal possessions, and they gave almost incredible mobility. For those reasons, the nation was jolted when the fuel demands of World War I brought people to the sudden realization that petroleum supplies were limited. Projections of domestic oil reserves (made for the first time in 1907-1908) indicated the sudden sharp increase in consumption could cause our wells to run dry before the end of the 1920s. *National Geographic Magazine* warned its readers solemnly in 1918 that they might have to give up their passionate love affair with the auto if we could depend on only U.S. petroleum production as a source of fuel. But—in the wonderfully optimistic spirit that characterizes that publication—the magazine went on to explain we had another, virtually limitless source of motor fuel on the verge of being tapped. Shale oil would save the day!

VII. 1918-1946: NATURAL GAS AND OIL COME OF AGE

Socio-economically, the next span of U.S. history was uneven. It was gashed in the middle by the Great Depression and climaxed by years of involvement in a second World War. Far from killing our long-term national confidence, however, adversity may have reinforced it. One energy theme emerged unmistakably to characterize the period: For the first time, liquids and gases came to assume the same order of importance as the solids, coal and wood.

Coal's ascendancy ended abruptly, even though we obviously were not running out of the fuel and its economics still made sense in the abstract. Although industrial coal consumption held reasonably steady, this was due to the increasing use of coke rather than to direct burning. The actual industrial *growth* in the U.S. turned largely to other energy sources, while the transportation sector, which had been responsible for coal's rapid growth in the first place, turned its back on the fuel. The rapidly expanding electrical generating industry (a major energy consumer while also serving as a "packager" of energy for easy end-use) hedged its bets as to fundamental source. That in itself was a major plus for liquids and gases.

Contrary to prediction, shale oil was *not* the answer to the threatened gasoline crisis—although near the end of this period (1944) the U.S. Bureau of Mines did open an experimental oil-shale mine and refining operation. Instead, the solution came from better methods of geophysical surveying and production that had not been foreseen—with results epitomized in the great new fields of West Texas.

As usual, the changing energy pattern involved more than just the direct substitution of one energy resource for another in the same use. Three technological developments which set the course of transition were widespread electrification, long-distance pipelines, and the replacement of railroads by autos and planes as the nation's favorite mode of intercity transportation. Once again, the Big Three—government, industry and labor—played significant roles in the change, although certainly no concerted "energy policy" was ever announced. In fact, the size and swiftness of the national shift toward liquid and gaseous fuel was barely noticed by a general public engrossed in more dramatic events of boom, bust, and battle.

Overall, this was not a time of rapid growth in our composite national consumption of energy. The historic rise in population tapered off and per capita use of energy in 1946 was only seven percent above what it had been in 1918. We became more productive as a nation with the energy we used; the amount of energy consumed per unit of real GNP declined rather steadily.

For this period comparative data are easier to develop. By this time, fuel statistics were tabulated both by industry and government, in part because of the various forms of taxation and partly because of concerns about accidents associated with fuel production (especially coal). The latter led to careful tabulation of the number of accidents *versus* mine output as well as accidents *versus* employment in comparison with other industries.

Surprisingly, Americans used progressively less energy during this period to reach measurable production goals (see Fig. 5). The real price of energy was dropping, but the same was not true of labor costs. Even during the depths of the Depression, the average real wage in the U.S. remained remarkably stable, never dropping below the 1918 level. By 1946, individual real earnings of the American work force were almost exactly double what they had been 28 years before; legislation and strong labor organization saw to that. One might have thought the consumption of energy in production would become more intensive—simply to reduce costs. Even practices which might otherwise be considered *wasteful* of energy could be defended under economic circumstances like these. Yet, in fact, a given amount of production came to require less energy-input—even though more energy was being expended *per worker*.

New ways of applying energy to what the nation chose to produce were more efficient. The evidence of their success was found in the almost miraculous output during the war years of 1942-45, which left the United States the richest and most powerful country in all of history.

While this kind of analysis may be acceptable on a grand scale in retrospect, it does not answer the question of how companies—or, even more to the point, *individuals*—reached the day-to-day

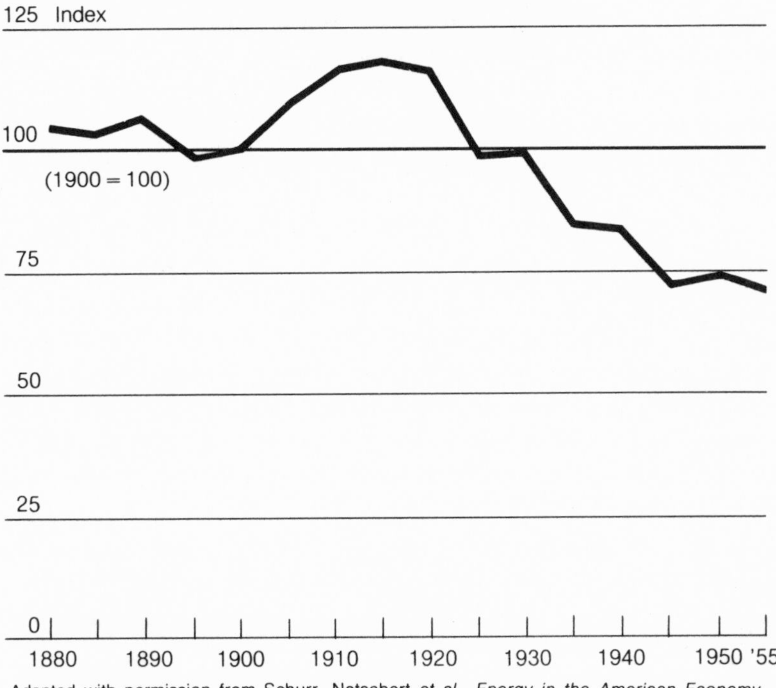

Adapted with permission from Schurr, Netschert *et al.*, *Energy in the American Economy, 1850-1975*, Copyright 1960 by The Johns Hopkins Press, Baltimore 18, Md. A Resources for the Future book published by the JHUP.

FIG. 5. Energy in this case includes mineral sources, hydropower and fuelwood, but not windpower or musclepower (although neither of these would be expected to change the general shape of the curve for this period). The total amount of energy used while producing a given output has stayed remarkably stable except for a steady drop between the end of World War I and the end of World War II. That decrease was probably due largely to more efficient methods of applying energy to production (including electrification).

decisions that added up to the new energy pattern. What happened to King Coal? Why and how did people begin to use so much oil and gas? What prompted electrification at this particular time, when it had not occurred more completely in the four decades since the striking success of Edison's light bulb?

In responding to the first question, the convenience and flexibility of motor vehicle transport explains a lot. Unlike trains, it offered door-to-door delivery. Furthermore, poor rail service during World War I had soured many customers, and the trucking industry got its first big chance when railroads concerned with long-haul freight began to refuse short-distance shipments of such commodities as milk and farm produce. After that, truck hauling grew quickly, surpassing rail freight as early as 1930. Even though gasoline shortages during World War II helped boost railway passenger miles back up to an all-time high in 1944 after a disastrous drop during the Depression, it was too late by then for railroads to recover, even as a means of personal travel. The family car had become a commonplace element of American life.

Perhaps—but *only* "perhaps"—railroads might have competed more vigorously in the freight market if they had switched to diesels earlier. Diesels were more efficient and cost-effective but, although they entered service in the mid-1920s, they were slow to replace steam. In 1946, seven out of eight U.S. locomotives were still steam-powered. Some of those used oil-fired engines but that phenomenon was limited essentially to the Western U.S., where ready access to petroleum enabled it to compete economically as a boiler fuel. For the most part, rail energy was still tied to coal . . . and the real price of coal was rising, on the average, while the real price of petroleum products, on which truck competition depended showed an enviable decrease.

The coal industry itself was caught in a classic squeeze. Labor unrest and stricter public insistence on mine safety standards boosted production costs, while shrinking sales also reduced return on investment. The tougher competition became, the more the original cost advantage available to the coal producers was eroded. The simultaneous surge in electric power generation might have turned the tide in coal's favor (as it was, it helped to keep the industry alive) if coal had remained the almost exclusive fuel for electric utilities. But it did not. Once again, there were regional considerations. In some parts of the U.S., easier-to-handle oil and gas were the "neighborhood" fuels that found a bonanza in the new demand for electric power, while hydroelectric dams dominated others.

As for the automotive revolution, Henry Ford's gamble of reducing the length of the work day and paying twice the going wage along a standardized production line probably would have failed if there had not been incredible public demand for the cars he turned out—and the first widespread introduction of installment buying plans by the mid-1920's. The romance between Americans and automobiles was in full blush and not even the dreary days of the 1930s could dampen its ardor. After the crash, the average American kept right on paying about the same retail price for gasoline, because new taxes nearly made up for the drop in base price. But—year in and year out—consumption per vehicle continued at a remarkably steady 11 or 12 gallons a week, even with commerce collapsed and joblessness rampant (see Fig. 6). People tightened their belts, but they hung onto their cars and kept driving them.

Although overall energy consumption showed a relatively small long-term increase during this period, there was a quintupling of the use of both natural gas and oil. The fact that the latter was tied overwhelmingly to the popularity of cars and trucks is shown in motor vehicle registrations (which skyrocketed from about six million in 1918 to nearly 35 million in 1941, before wartime mobilization shut off civilian production) and in the way refinery output shifted. A barrel of crude oil can be processed in various ways, depending on which of the products of fractional distallation are to be emphasized. In 1918, when kerosene was still in widespread use as a lamp and heating fuel, the average ratio between gasoline and kerosene in the output of the nation's refineries was two to one. By 1946 the process had been modified to produce eight gallons of gasoline for each gallon of kerosene from essentially the same crude oil. Auto use helped to boost demand for natural gas too, because in those days about 10 percent of all natural gas was used in the manufacture of carbon black for rubber tires.

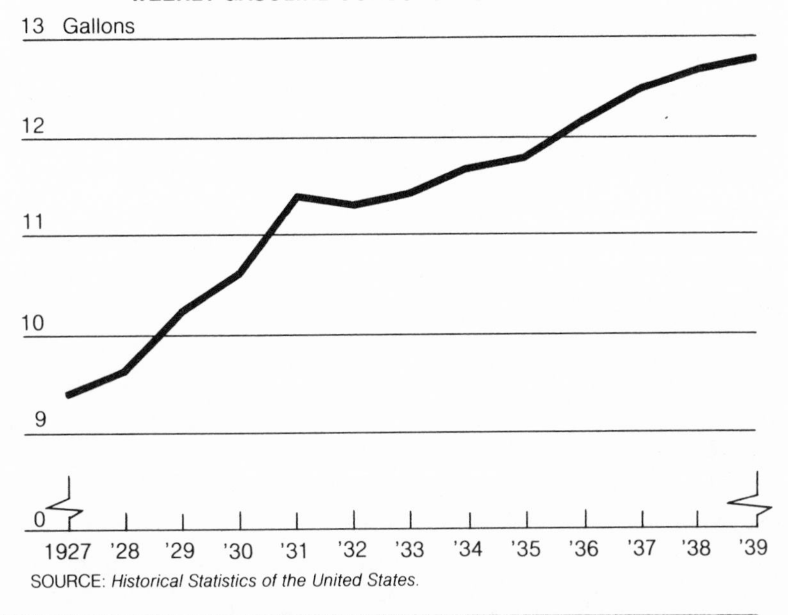

FIG. 6. The Great Depression caused no more than a slight pause in the increase of gasoline use, while motor vehicles were replacing trains as this country's favorite form of transportation. The number of vehicles increased during this period, of course, but so did individual fuel consumption—as shown by dividing the total usage of highway motor fuel by the number of autos, buses and trucks registered in the U. S.

Electrification was another counter-cyclical trend through much of the Great Depression, thanks in large part to government subsidy. The Boulder Canyon Act of 1928 had ordered Federal sales of electricity to be generated at the new dam site, and this led to the development of the first really long-distance and high voltage transmission lines. Power was delivered across the desert to Los Angeles as soon as Hoover Dam was completed eight years later.

Of course the same technology could be and was applied elsewhere. Franklin D. Roosevelt's New Deal pumped money eagerly into projects like the Tennessee Valley Authority, the Bonneville Power Administration and the Rural Electrification Administration. Although he did not refer to it as an official "energy policy," FDR apparently was trying to make as much energy as possible available to as many people as possible at the lowest prices possible by introducing generating entities with low-cost or free government capital and freedom from taxation.

Electric lights and indoor plumbing became the twin criteria of minimum modern housing standards. The electrification program created jobs and it offered a visibly "better way of life" to millions of Americans who otherwise had little cause for cheer. The use of electricity spurted four times as rapidly as the country's overall consumption of energy. The fastest growth, by far, was in residential/commercial use rather than industry. Hydropower was the favored Federal vehicle, and there were times during this period when 40 percent of America's electricity came from dams.

At the same time, the steam generating equipment operated primarily by investor-owned utilities was improving in efficiency, and the expanding market for electric power made possible striking economies of size. There was a snowballing effect. Electric rates dropped everywhere, and—as in the case of automobiles—electrification changed from the category of a rare luxury to a convenience taken for granted. In 1918 only about one-quarter of all U.S. dwellings and barely one farmhouse in a hundred had electric service. By 1946, electric power was being used in 85 percent of all American homes—even including a majority of farms. Without widespread electrification, radio's enormous impact, with news, entertainment and direct oral reassurance from the Chief Executive, could never have been achieved. For that matter, the movie industry could not have emerged either.

With gas losing its marketability as a lighting source, the emphasis of the supplying industry had to shift to heating—particularly in the form of home kitchen ranges, and later in furnaces as well. This meant more bad news for coal, because one of the strongest new selling points for gas was how clean it was to use. Residential coal bins became an annoyance if not an outright embarassment and a cry went up for "smokeless cities"—which advertising indicated would also require industry to switch away from the dirty black fuel as soon as it could.

At about this time, natural gas began to replace the various types of coal-based manufactured gas that had been in general use across most of the country. The switch was made possible in large measure by improved pipeline delivery techniques, and better ways of storing gas in reasonably large quantities were developed. A given volume of natural gas provides far more heat when it is burned than the old synthetic gases. The price of gas at the wellhead was extremely low, since it was often regarded as a "waste product" from oil production, so consumers tended to get a break even after utility rates were adjusted to cover long-distance transport costs. Natural gas became a truly national fuel for the first time, although its primary market was in industry and much of this was still limited to what is called "field use"—providing energy for drilling operations and nearby refineries. At least this was better than "flaring" it—the common earlier practice of burning off the gas that came from oil wells just to get rid of it safely.

The most dramatic government contribution to the eventual natural gas network was the "Big Inch"—a pipeline two feet in diameter that stretched more than 1,250 miles from the Southwest to the Northeast after its completion in 1943. Actually, it was converted to carrying gaseous fuel only after World War II. Originally it carried oil—at a time when the U.S. was hard pressed to keep its overseas fighting machine fueled and gasoline for civilian use was rationed.

Americans had no alternative but to cut back on their driving during those wartime years, but by that time the nation had already developed a substantial highway complex. This had been the product of an irresistible public campaign for roads earlier in this period, beginning with the "Lincoln Highway Association", which collected funds directly from the public for the building of a through road from New York to San Francisco. Donations financed what were called "seedling miles" (short stretches of good pavement along various parts of the route to develop further public support for better roads). Heavy pressure was exerted on Congress to take responsibility for a national road system that would permit quick and comfortable cross-country travel unaffected by state lines. The 1916

Federal-Aid Road Act set the pattern of national appropriations to be expended by states for highways required to meet certain common standards. A principle was also established early that highway users should bear the cost of roadbuilding and maintenance, either through fuel taxes or direct assessments on commercial haulers. As costs rose, the completion of the Pennsylvania Turnpike in 1940 opened the modern era of "expressway" toll roads. Americans took proprietary pride in their unsurpassed highways, and it is no wonder one of the most talked about features of the New York World's Fair in 1939 was the General Motors pavilion. It gave visitors a preview of the concrete lacework that would cover the countryside in our World of the Future.

Aside from the fact that organized labor provided overwhelming political support for the policies of the New Deal, its most obvious effect on national energy habits was a negative one. The coal industry was sick, and so were the railroads. Violent strikes made the illness worse. As for the oil industry, its malady was different. The corruption uncovered in connection with production arrangements for the Naval petroleum reserves at Teapot Dome was surely as much of a reflection on morality inside government as in private enterprise, but this did not prevent serious consideration during the mid to late-20s of direct federal involvement in the petroleum industry. The agitation for such a move was based largely on the fact that petroleum reserves were clearly going to become more important to both our economic and military security, but a combination of industry self-regulation and a more orderly supply picture averted drastic intervention. The threatened shortage was replaced for a while with overproduction and cutthroat competition. Still, the industry had been put on notice by the new Federal Oil Conservation Board and its own studies that there would be a continuing need to uncover and develop new reserves as well as to maximize the total useful production from existing fields. By 1938, U.S. drilling reached a record depth of 15,000 feet. By the mid-40s production platforms had moved offshore into the Gulf of Mexico.

Even in the late 1920s and early 30s, U.S. geologists had searched farther offshore—going all the way to Latin America, the Persian Gulf and the Far East to finance and direct the development of what seemed to be a new supplementary source of "easy oil." The expropriation of foreign oil holdings in Mexico in 1938 left American petroleum companies apprehensive about the possible recurrence of such a problem, however. Subsequent agreements with foreign governments invariably involved a form of multinational partnership.

Much of the petroleum development that had taken place on other continents resulted from British and Dutch initiatives, some of which started before World War I. U.S. entry into those areas was greeted with a certain hostility. Nevertheless, the State Department insisted on a sort of "Open Door Policy" in regard to overseas oil development and U.S. companies gradually expanded their production activities around the world. A brief effort in 1928 to press American membership in an international cartel to allot market shares and maintain prices in the face of an impending world glut dissolved when it became clear anti-trust laws in this country prohibited agreements among companies to curtail domestic production. It should be remembered that throughout this period the United States was a fairly significant *exporter* of petroleum.

Aviation blossomed technically during these decades, but was a minor factor in total energy consumption. By 1941, annual production of civilian aircraft in the U.S. amounted to only a few thousand. Only a few hundred of the planes in service were regular commercial carriers. There was great skepticism about the economic feasibility of large scale, long-distance transport planes because of the calculated horsepower requirements, and that helps explain the popular success of airships—dirigibles—for somewhat over a decade in the 20s and 30s. By getting an assist from lighter-than-air gas, a giant like the *Hindenberg* could use four diesel engines to carry some 50 passengers in splendid comfort across the Atlantic—moving more slowly than planes, of course, but still at approximately four times the speed of a steamship. In the U.S., however, dirigibles were basically a monopoly of the armed forces. For that matter, the military ordered about 35 percent of the more than 50,000 airplanes built in this country between the world wars.

All-in-all, this was a period when there was great public faith in the ability of technology to overcome virtually any obstacle. In the 1930s there was even a movement to launch a "Technocracy" in which scientists and engineers would take political control and solve all the nation's domestic and international problems promptly by such grandiose measures as diverting the Gulf Stream. FDR himself was intrigued by the natural energy displayed in huge tidal changes near his retreat at Campobello, and he tried unsuccessfully to institute a federal project to tap the water movements there as a source of useful power. At Grandpa's Knob in Vermont, an electric company installed a wind turbine whose generator output was large even by today's standards for a windmill—1.25 megawatts. But the fact is that such exotic schemes were unnecessary and uneconomic then because conventional energy was cheap and plentiful. If the nation could maintain the same link between energy consumption and productive output, there seemed no reason why succeeding generations should face any resource crisis either.

By the end of this period there was also the strange and exciting new prospect of nuclear energy. Looking back now, it is understandable why its development took the American public so thoroughly by surprise in 1945. After an earlier flurry of attention to the possibilities of nuclear fission and fusion, U.S. physicists had adopted a voluntary agreement well before we entered the war not to publish further results of nuclear studies for fear of helping the Nazis develop a bomb. As late as 1938, E. O. Lawrence, who developed the cyclotron, wrote that the chances of turning matter into energy for the purposes of generating power were no brighter than those of using heat from the oceans—another purely theoretical possibility. Thus impressed by an "unexpected breakthrough," journalistic projections of the future in 1946 naively concentrated on full nuclear conversion potential rather than the realities of engineering. A phrase often repeated in articles at that time "explained" that the energy in a glass of water was enough to fuel an ocean crossing by a big liner.

Believable? Of course! Had not history already revealed we could do anything we really set our minds to?

VIII. 1946-1974: ENERGY SPRAWL

Superficially, U.S. energy statistics for the period between the termination of World War II and the declaration of the Arab Oil Blockade in late 1973 resemble those of 1880-to-1918. On the strength of the postwar baby boom, population growth picked up—resulting in a net population increase of about 50 percent. Energy use per person grew at an almost identical rate, so total energy consumption more than doubled. Again, real GNP paralleled rising energy use. Each of the two periods closed amid concern whether domestic supplies would be able to meet expanding demand.

But there were differences. For one, the nation had used the decades prior to 1918 to tool up as an industrial power. By the late 1940s, the U.S. was already a mechanized, automated, productive colossus on the threshold of the computer revolution. Enormous quantities of the capital equipment and sophisticated technical machinery turned out in this period were for export. Instead of building our own capacity exclusively, we undertook to rebuild or develop industrial capacity throughout much of the rest of the world.

At home, more output was beamed toward the ultimate consumer market. There is no doubt that in the first two decades of the 20th Century, the U.S. was a generally comfortable place. But now—half a century later—we had became affluent. Most Americans enjoyed the sensation. There was also a crucial difference in where our potential energy shortages lay. The imminent problem with oil reserves at the end of World War I involved a fuel we prized because it had made possible the beginnings of our mobile economy, but not a fuel we had come to depend on very heavily in comparison with other sources. Crude oil then contributed scarcely 10 percent as much to our energy consumption as coal. In contrast, the 1973 oil embargo hit us after we had converted our economy into one that derived almost half of its energy from petroleum and another 30 percent from natural gas. We were far more vulnerable.

Demand repressed by the World War II mobilization exploded in the following years. Nowadays we may not be greatly impressed to be told that more than 200 million motor vehicles were produced and sold during the 1946-1974 period—even though this represents a minor manufacturing miracle. But it is also part of the overall story that well over 100 million cars and trucks were *junked* during that same time. We reached the stage where car-owning families approached 90 percent of the population and about one family in three had more than one. We drove faster and scrapped cars more quickly. Thus, affluence took a double toll in energy consumption: People were using nearly 10 gallons of gasoline a week for every man, woman and child in the country and were also using a significant amount of that resource in the production and distribution of goods, from cars to cardboard cartons, destined for discard long before reaching what might have been considered the end of their "useful lives."

Much of U.S. social history for this era revolves around the motor vehicle. People stopped riding trains. By the early '70s, railroads were supplying less than half of one percent of our national transport in terms of passenger miles. Streetcars gave way almost universally to buses, and all public transit decreased by about 50 percent. Airlines accounted for far more personal travel than railroads had ever handled at their peak but did not come close to competing with the motor car for sheer transportation volume. At the time of the Arab Oil Embargo, the annual intercity travel in this country broke down very roughly like this in passenger miles: railroads, 10 billion; air, 100 billion; private car, more than *one trillion*! And local auto traffic may have matched that figure.

Considering what can be delivered and how much energy it takes to get it there, it is hard to find fault with motor transport (or with air travel, for that matter). As a matter of fact, the transportation share of U.S. energy consumption actually dropped slightly between '46 and '73. The fateful change that took place, however, was that coal (a plentiful domestic resource) virtually disappeared from the transportation sector. The railroads' use of petroleum-based fuel (both oil and diesel) remained about the same during this period, but their consumption of coal dropped from more than 100 million tons a year to essentially zero. The difference in transportation's slice of the growing energy pie was made up by our new daily diet of over six and a half million barrels of gasoline, about a million barrels of jet fuel, and more than half a million barrels of motor fuel for diesel trucks.

It was certainly no secret that the nation had come to depend on imports of crude oil to satisfy the increasing thirst for all petroleum products. By 1958, the United States had begun to import more energy resources overall than it exported. By 1970 domestic oil production reached its peak and started to decline. Two years later, in a chilling preview of things to come, the country recorded its first trade deficit since the depression year of 1935 and only the second such annual deficit since 1888.

This was really the first era in which the truly global transfer of energy resources became a major component of world trade. Pipelines crossed borders and supertankers criss-crossed oceans. In the spirit of those times, it is easy to understand why these developments aroused little public concern. Supermarkets in Middle America had become accustomed to satisfying their customers' taste for California and Florida oranges. Even global distances were discounted. It made economic sense (in fact, it was a sign of personal frugality) to buy Volkswagens from Germany and electronic equipment from the Far East. From a dollars-and-cents perspective, it seemed logical that those areas where petroleum could be produced most cheaply ought to specialize in providing it to everybody else; after all, we all inhabit the same planet.

The average price of crude oil roughly doubled within two years after the end of World War II. From then on, however, the real price of crude tended to remain fairly constant—even, after about 1960, to decline. In large measure, this was a reflection of competition from lower production costs overseas, where there seemed to be loads of "easy oil" (that is, oil easy to find and produce). By 1950, rapidly expanding exports from the Middle East just about forced U.S. oil off the European market but our demand at home easily took up the slack. Major U.S. oil companies were heavily involved in various forms of cooperative production abroad, so they accepted these foreign sales developments without much complaint. Yet, importing cheap oil from the Persian Gulf into the U.S. raised some fears that domestic markets for domestic crude might shrink disastrously. Voluntary restrictions failed, and executive action by President Eisenhower in 1959 set up a system of petroleum import limits and allocations. It's ironic that this move was roundly criticized by some as unfair favoritism to the domestic oil producers, but that a somewhat similar suggestion by President Carter 20 years later would be hailed as an indication our country meant business in its effort to regain a measure of energy independence.

European supplies of Middle East oil, in fact, had been cut off temporarily after the Anglo-French invasion of Suez in the mid-50s, but at that stage the U.S was able to step up its exports on an emergency basis and many regarded the speculation about new political interruptions as unrealistic. Persian Gulf prices remained low, dipping to $1.20 a barrel in 1970, and it was only when worldwide surplus oil production capacity melted away (i.e., when U.S. production stopped rising), that OPEC had the ability to boost its charges unilaterally.

The pieces of the new energy picture now were all at hand and fell into place suddenly. As domestic output dropped and domestic requirements increased, U.S. imports of crude oil and refined products soared from 3.4 million barrels a day in 1970 to 6.2 million (more than one-third of all our nation was using). That is where imports stood in 1973, when another Arab-Israeli War precipitated an embargo against the West and the start of a dizzying upward oil price spiral.

The fact of American affluence is worth repeating, because it affected energy usage in a great many ways. People were eating more meat and poultry, a demand livestock feedlots and "chicken factories" met by energy-intensive production methods. Frozen foods had become a part of our lifestyle too. Unemployment had gone well above six percent on two occasions, 1958 and 1962, but for about half of this period it stayed below four and a half percent—even though the official national labor force was expanding steadily to include larger numbers of females than ever before in history. By contemporary standards (and certainly in comparison with the period before World War II) this was a time of "full" employment nationwide. Steady paychecks kept consumption of all goods and services relatively high. Because all commodities and most services are based at least in part on energy in some form, energy consumption also was high.

Although there were still disturbing inequities on a racial basis, the real income of individuals and families rose markedly. Notwithstanding a solemn "War on Poverty," the poor remained with us; but the *meaning* of "poverty level" (and even "middle class") changed. Using 1967 purchasing power as a common yardstick, the national fraction of families with real incomes of less than $7,000 dropped from more than three-fourths in 1946 to barely one-third by 1973.

If social conscience motivated this transformation, productivity helped make it seem reasonable and natural. The value of U.S. manufacturing output per worker-hour doubled, although the rise in wage rates for non-farm industry rose even more sharply. That meant unit labor costs were increasing—while the declining real price for most fuels made energy a smaller cost factor even when more resources were consumed. From a purely economic standpoint the long-standing incentive to substitute labor-saving machinery and methods was further reinforced. Nowhere was this more obvious than in agriculture. Over the period, farm output per man-hour went up almost fourfold, the most dramatic overall rise in U.S. history. Simultaneously, of course, the actual number of farmers dwindled. By the early 1970s, it took less than half as many agricultural workers to feed a U.S. population half again as large as it had been a generation earlier—with more than six billion dollars worth of farm produce on top of that for export.

The additional energy used in agriculture was based almost entirely on petroleum. In fact, the only two purposes for which this country was still using coal in any appreciable quantity were metals production, in the form of coke, and the generation of electricity.

The unit price for a kilowatt hour of electricity declined steadily during most of this period and its use expanded accordingly. The fastest growth came in homes and businesses, where, around 1961, air conditioning commenced a remarkable upsurge. This was especially significant because electricity differed from many other energy forms in the inability to stockpile it. Large-scale storage of electric energy is very costly. That means enough generating capacity had to be made available to meet peak demand—even if that maximum requirement is confined to afternoons during a couple months of the year, when air conditioning systems are turned on full blast. A broader "generating mix" began to evolve—ranging from capital-intensive but operationally efficient "baseload plants" (which hum along steadily around the clock, month in, month out) to various sizes of quick-responding units (which might not be practical for continuous operation, but could supply peak bursts of power). At the same time, more extensive transmission networks had to be planned and put into place. They would be needed to make such a system work within a single utility area in any event, but they could also take advantage of time differences in load peaks among neighboring systems by interconnecting them.

At almost the same time, however, another vigorous complication was developing. A new citizen activism expressed itself partly in the "environmental movement," but it was really broader and deeper than that alone.

The period 1946-1973 was an era of affluence, but not a time of calm. During about half of the period, U.S. armed forces were involved in combat overseas, although neither Korea nor Vietnam involved a formal declaration of war. Constitutional precedents were overturned and statutes of long standing were upset or redefined in the interest of human rights. The "establishment" was challenged in every way imaginable. The period closed with the resignation of a Vice President and then a President.

Anxiety about the environment was surely no novelty in this country when Rachel Carson's *The Silent Spring* was published in 1962. Yet, a social psychologist might easily argue that the seeds of concern planted anew by that book found unusually fertile ground in a country whose affluence gave its people more time to consider such problems. Groups of "environmental activists" that sprouted and spread during the following decade were undoubtedly nurtured by general feelings of societal guilt, restlessness, and an amorphous idealism that was willing to let "business as usual" grind to a halt while what they conceived as "wrongs" were being set "right."

The Wilderness Act (1964), National Gas Pipeline Safety Act (1968), Federal Coal Mine Health and Safety Act (1969), National Environmental Protection Act (effective January 1, 1970), Federal Clean Air Act (1970), and Federal Water Pollution Control Act (1972) are some of the highlights among important new legislation that affected the production and use of energy. To cite just one example, electric utilities began a sweeping conversion of baseload coal plants to the burning of oil. Even at that point, oil was the more expensive fuel, but its use made it possible to meet new regulations on emissions from the generating plants' stacks. In only three years (1970-1973), the amount of oil going into the generation of electricity in this country rose by 685,000 barrels a day. To help tighten the screw, a price ceiling on natural gas that crossed state lines was cultivating a shortage of that alternative fuel.

A Supreme Court ruling in 1954 in the case of *Phillips Petroleum v. Wisconsin* had extended the Federal Power Commission's jurisdiction from gas pipelines crossing state borders to the wellhead prices paid producers to supply such pipelines. Two years later both houses of the U.S. Congress voted to deregulate natural gas, but a flap over industrial lobbying caused President Eisenhower to veto the legislation. Because prices were then pegged at artificially low levels, natural gas became a most attractive fuel for industry; but domestic production peaked in 1972. Interstate customers could not compete with markets for natural gas within the producing states, where prices were free of Federal control and simply followed free market trends. So far as electric utilities were concerned, economically recoverable reserves (based on regulated prices) seemed woefully small in the face of simultaneously rising demand for home heating and petrochemical uses.

Meanwhile, civilian nuclear power plants also were encountering their first organized roadblocks. In a military context, the nuclear propulsion of submarines had never faced any true competition in terms of performance. The atomic subs—and surface ships, to some extent, too—could do things conventional marine powerplants had no way of matching. But converting nuclear reactors to peaceful use raised new questions of cost effectiveness when compared with conventional plants. No one who understood the technology had ever echoed AEC Commissioner Lewis Strauss' boast that nuclear steam sources would make electricity "too cheap to bother metering." It was obvious that economical nuclear generating systems would need large individual outputs to spread the inevitably big capital outlay over a high baseload capacity of electricity with relatively low costs for fuel, operation and maintenance. Nevertheless, it was by no means obvious which type of reactor would be best for civilian power plant applications, so a few small "demonstration" units were built in various parts of the country to test different designs. As it turned out, "light water reactors" that were similar in concept to those developed by Admiral Hyman B. Rickover's team (although the civilian versions used much lower percentages of fissile uranium in their fuel) eventually proved most successful. Not until 1970 or 1971 did commercial-sized plants enable nuclear power to surpass wood in its total energy contribution to the nation, but a very large backlog of orders for follow-on plants seemed to indicate that the future of nuclear energy was secure. A projection of some 1,000 reactor units by the end of the century was made quite seriously. However, that was before a succession of objections by intervenors in the plant licensing process began to be voiced. The vogue of complaints changed over time; charges of radioactive emissions during normal operation gave way to fear of undesirable thermal effects, and later to questions about the possibility of catastrophic accidents. Responses to each objection as it arose failed to stem a steady stream of protests. The only thing that stopped (in unfortunate coincidence, just about the time of the Oil Embargo) was the placing of orders for nuclear power plants by U.S. utilities.

The existence of vigorous and occasionally violent protests of various kinds might suggest individualism was ascendant in the United States of 1946-73. To some extent it was, but quantitatively this trend was certainly overwhelmed by a counter-move toward national homogenization. Easy mobility and instantaneous communication (a sort of "mobility of ideas") were the blending factors. The automobile, airliner and television were the principal instruments.

Several international expositions were held in this country during the period, but none seems to have carried the special impact of those in earlier days. Instead, Disneyland, Disneyworld and countless imitators of the "theme park" idea were open year-in and year-out—as part of a bright new part of the American scene christened the "leisure industry." Entertainment, rather than enterprise, was its keynote. Even the 200th anniversary of the Declaration of Independence failed to produce a successor to Philadelphia's famous Centennial Exposition—although tens of millions of Americans were able to watch coast-to-coast celebrations of July Fourth in living color from their homes, and hundreds of thousands in the nation's capital succeeded in producing the premier of all traffic jams after watching fireworks on the Mall.

The U.S. had essentially a rural population for most of its history. Briefly—approximately for the era between the two world wars—this was a basically urban nation. Since then it has become dominated by suburbs. Cities like San Francisco, New York, Baltimore and New Orleans retain their individuality, yet their suburbs are remarkably alike. In dozens of ways, suburban living and "suburban psychology" have tended to be energy-intensive. The phenomena of daily commuting, fast food outlets, conspicuous consumption, throwaway products, universal emphasis on petroleum-based synthetics, and "drive-in everything" have been explored so thoroughly and are so likely to be fresh in every reader's perception that no new litany seems necessary.

The decaying cores of our cities were unintended casualties of our national energy sprawl—innocent bystanders, caught in the crossfire, so to speak. Another memorable aspect of the era that had only an indirect connection with energy habits at the time was the exploration of space. The eventual impacts on energy development were not perceived at first.

Space propulsion is a minute user of energy in the overall scheme of things, but spinoffs from the space program are legion in their energy effects. Transistors and integrated circuitry, vastly improved photovoltaic cells, fuel cells, thermoelectric converters, lightweight chemical batteries, cryogenic systems, and a vast array of new materials and joining processes all relate to energy. Furthermore, those are only the "things" of space development, quite apart from new capabilities. The latter include a better understanding of geological processes, the improved prospecting that results from surveying earth resources by satellite, intercontinental and interplanetary video communication, and the refinements in navigation that make remote areas of the globe much more accessible. The irony is that the nation did not really have domestic energy problems on its collective mind when most of these breakthroughs took place.

As the Age of Affluence drew to a close, the shock of having cheap energy withdrawn suddenly was presaged by a more modest and gradual domestic price escalation that began in 1970. A downward trend in the real price of petroleum was interrupted. Restrictions on the use of high-sulfur coal caused the average cost of that fuel to rise. A steady decline in the average cost-per-kilowatt for most users of electricity was turned around for the first time in nearly half a century.

Industry—whose overall share of all energy consumption dropped from nearly 50 percent to less than 40 percent between 1946 and 1973—was sensitive to the impending price storm during those last three years. Starting in 1970, the average amount of energy consumed for each unit of industrial output dipped annually.

Personal users of energy reacted differently, however. In 1946 the household-and-commercial sector of the economy was using one-quarter of all U.S. energy; by 1973 that sector's share was more than 35 percent—an enormous increase when one considers the growth in the size of the whole energy pie.

The man in the street was totally unprepared for either a short-term energy crisis or a long-term national energy problem. Perhaps the saddest feature of this entire segment of U.S. energy history is that it left Americans ignorant of the basic structural changes that had taken place in both supply and demand. Man had left footprints on the lunar surface. The penchant for quick travel, fast living, and immediate answers offered little patience with any problem that threatened to drag itself out. Regardless of what they thought about President Nixon in other contexts, many Americans probably trusted the feasibility of his "Project Independence," which would set everything right by 1980. People did not realize how preposterously optimistic that analysis was.

IX. 1974-2004: UNCERTAINTY AND ADJUSTMENT

Even though historic cycles appear to exist, events of the past cannot pinpoint events of the future. Circumstances are never repeated precisely, and time even modifies the pattern of human responses in a continuing society.

The "periods" into which this history is divided were chosen partly as a matter of convenience in exposition, so it may be an artifice to select one of them as a precursor to the one the nation has started most recently. Nevertheless, there are hints that the three decades between 1974 and 2004 *could* be similar in certain characteristics to the period between 1845 and 1880. Motivating circumstances are different, but the present period has started in the direction of another "time of turmoil and transition." Of course, no one can be sure where it will lead.

Although another full-fledged Civil War is unthinkable, the ideological battle between "expansionists" and the "small-is-beautiful" forces could tear the economy apart unless some effective consensus of attitudes can be developed.

In an Orwellian coincidence, the election in 1984 might be the moment of crisis. Short-term and regional difficulties with various aspects of the U.S. energy picture are now expected, and they promise to be more severe than any that have preceded them. They could come to a head by 1982 or 1983, and energy policy might easily be a more pervasive element of the next presidential campaign than in 1980. The damaged reactor at the Three Mile Island nuclear plant could be ready for renewed operation around that time, and its status could become a focal point for debate.

Regardless of whether it seems pessimistic or optimistic, much of what has been written since 1974 about America's energy future takes too narrow an outlook. Some projections are based almost entirely on technological capacity, as if to suggest anything worthwhile that *can* be done by applying national resources *will* be done. Some energy futurists, on the other hand, concentrate only on economic analysis, implying dollars-and-cents logic inevitably will guide human behavior. Still other types of energy forecasts can be categorized as "wishful thinking" on the part of the individuals and groups responsible for them. It is easy for anyone to line up an impressive array of arguments about why developments one favors will take place or why trends that one dislikes must peter out.

Ideally, *all* of the major technological, economic, philosophical and political prospects should be considered in trying to look ahead to the end of the current energy epoch. Realistically, most projections need to be hedged. The possibilities raised in this chapter of energy patterns that might develop are all based on an analysis of our present situation and its similarities to the past.

What *are* the "lessons of the past"? One seems to be that new energy sources are often linked intimately with new ways of *applying* energy. In cases where one fuel is substituted directly for another, *availability* has usually been a factor—involving the distinct elements of final cost and deliverability to the end user. Most U.S. energy shifts in the past have been the result of free choice in the market place, although a few changes of relatively short duration have been based on genuine (though relative) shortages. Government has often steered developments—sometimes indirectly, if not unconsciously. The industrial-commercial sector has stressed technological progress as a labor-sparing device—with *some* form of non-human energy always having been available thus far to satisfy new demands. Above all, however, the past suggests that the future will hold surprises in technology and human behavior.

Some current statistical trends involve enough built-in inertia to venture fairly firm guesses about the next 25 years. For instance, population growth in the United States will probably be the lowest in our history—making us a nation of perhaps 265 million by a few years after the start of the next century. Without venturing too much detail about *sources* of energy, one might also tentatively accept what seems like a balanced projection of total energy demand made recently by the Washington think tank Resources for the Future. It implies a demand of about 120 quadrillion Btu's (equivalent to roughly 60 million barrels per day, if it were oil alone) by the year 2004. That would mean an increase in per capita consumption of only about 25 percent—small by historical standards but above what some ardent conservationists think it can or should be.

Because the greatest opportunities for conservation now on the horizon are in transportation and residential use, it might be surmised—with a bit less certainty—that the industrial sector of the national economy will consume a larger percentage of the energy we use by 2004 than it does now. This makes sense because energy is likely to be more expensive in the future and its extraction is likely to absorb more energy in itself. In the case of the remaining natural gas reserves in this country, for example, about 20 percent are in far-off Alaska, 20 percent offshore, and another 20 percent presumably will have to be tapped by very deep drilling.

All of this might be tied too closely to approaches of the recent past, however. Couldn't life styles be transformed? What about technical innovation? What of the possibility of societal upheavals if failure to implement a reasonable energy policy leads to a truly cataclysmic crisis?

It seems safe to assume a sizable amount of U.S. energy by 2004 could come from what now are considered "unconventional sources." By itself, however, that is not terribly specific. After all, in a hunting and gathering society, *agriculture* would be an "unconventional" food source; yet farming seems like second nature once it starts. The same ambiguity applies to energy sources at any stage. It is really too early to be sure what detailed form our next transition in energy technology might take. Synthetic methane, from coal or cow manure, is not "natural gas" as we have known it but its chemical composition and burning characteristics are similar. "Shale oil" is not really oil, but can be processed into a very close approximation of familiar petroleum products. Thus, certain aspects of the technological change beginning to take place might be almost undetectable by the average individual consumer, except in retail prices.

In part, technological change is based on which areas attract the most research interest. Such interest, in turn, depends partly on what a culture seems to need technologically. Perhaps that provides some other clues to the future.

Solar energy, wind turbines, and small-scale hydropower all require improvements in energy storage techniques if they are going to be more than marginally useful. Geothermal energy, solar heating, certain exotic schemes to tap wavepower or temperature gradients of the ocean, and even the more efficient application of nuclear reactors would be affected profoundly by successful new methods of utilizing small temperature differences and relatively low operating temperatures. None of this sounds very dramatic and exciting, but it is precisely this sort of "horseshoe nail" that could decide which way our energy battle is going to tip.

All energy sources depend on availability for their value. To a large extent this attribute has been closely related to transportability. Here is another case where research and development could make a difference. Every form of energy now used could stand improvement in distribution and delivery systems.

Big breakthrough possibilities should not be overlooked, either. Over the history of technology, there has been one striking innovation after another: the steam engine, electric power, nuclear reactors. It would be foolhardy to build an energy policy around the assumption that something extraordinary like those will turn up again within the next generation. Yet it is also unrealistic to ignore the possibility that one might. How can anyone rule out the chance that somebody will come up with a way of splitting neutrons to release the binding energy within them? (Remember the "indivisible" atom?). Or, take another possibility: The mysterious force within quasars may or may not turn out to be much more remote today than the thermonuclear reaction on our sun was when *that* was first described in 1920. Where might research on *earthbound* "quasar reactors" be by 2004?

Even a fairly thorough technical understanding is not the same as useful application, however. Fuel cells and photovoltaic conversion were recognized in the 19th Century but not considered on a practical basis until rather recently. Economic feasibility is also part of the story, and reaching the economic threshold also depends on social encouragement. Translated into today's terms, that latter phrase means *government policy*.

Government involvement in U.S. energy patterns can not be expected to decrease in the future, short of a total political upheaval. The federal government owns too large a share of remaining natural resources for that to be possible. It is too large an employer and consumer itself. Its control of the money supply, not to mention its ability to affect energy habits directly through regulation, is prodigiously important. And, on top of all these reasons, energy supplies are clearly related now to national security—an obvious government responsibility.

This does not mean the forces of free enterprise must decline. If government were consciously to *restrain* its regulatory and proprietary role in deference to market forces, that would still be a positive government *action*. Postponing action or refusing to take a stand at all on some aspect of energy development is similarly a form of government decision for which historical precedents exist.

A more complex question to explore is how political processes over the next two and a half decades will mold government policy, and here is where more speculation is involved. Attitudes in the "Sun Belt" could easily dominate the national future. The 1980 census probably will show half of the net population growth since 1970 has taken place in only three states—Florida, California and Texas. That may affect both Congressional composition and Presidential politics. Government and community leadership will also be passing to a new generation, one with no recollection of America before its period of "energy sprawl" and thus no ability to make personal comparisons with earlier circumstances. Finally, affirmative action programs will change the face of leadership in all fields too, giving more clout to women and members of racial and ethnic minorities previously underrepresented. American Indians, in particular, could have a great deal to say about energy policy because of their ownership of, occupancy of, or claims to lands containing energy resources. What all this might mean is open to controversy.

Even the groundrules in the political decisionmaking process itself could be modified. A cursory mental review of history indicates how things have changed through successive broadening of the electorate, direct election of Senators, shortening of the "lame duck" session between elections and inaugurations, and other political changes. They affected energy habits in the past, and comparable changes could do so again. For instance, with tension between the executive and legislative branches continuing, it is not difficult to envision a shift toward a parliamentary form of government—perhaps after one or two experiments at the state level. This might either speed up energy initiatives or introduce new instabilities in policy. On the other hand, the pendulum could swing the other way—toward firmer exercise of independent executive power, perhaps through the mechanism of a longer single term for the President and the emergence of some magnetic, persuasive, and forceful personalities.

It would seem contrary to the fast-moving tides of change if a major new political party did not appear between now and the end of the century. If it does, what will its platforms and actions produce in respect to energy? How will they affect big business and labor?

Worldwide events will bear on U.S. energy policies too. They always have. Operating as it does in a global context of supply and demand, the United States will be affected inevitably by the growing energy demands of the Communist economies and the Third World nations, both of which have increased even more rapidly than the

United States' since the end of World War II. Between then and the time of the 1973-74 Oil Embargo, developing countries doubled their share of commercial energy consumption—climbing roughly from five to ten percent of whole-earth totals. By 2004 they are projected to account for fully one-third of annual energy use. Their demands and ours must compete if energy supplies are limited.

Political volatility within various energy-exporting countries is another feature of uncertainty. There are many ways U.S. patterns of energy planning could be disrupted by events in another part of the globe over which the country has little or no control. More ominous is the simple historical fact that it will be an exception to the rule if the U.S. survives the next few decades without involvement in a war. A major depression would distort the energy picture too, as could some natural catastrophe such as city-wrecking earthquakes.

Whatever happens, what cultural attitudes and philosophical underpinnings will determine people's reaction to either the anticipated or the unexpected? Once again, it is largely a guessing game—because of sharply conflicting views within the energy policy debate. Socio-economic reforms in the past have proved practically irreversible, so it is hard to imagine the U.S. will reduce its commitment to relatively high real wages, environmental protection, and rules to safeguard public health and safety. At the same time, the nation is probably too dependent upon the benefits of an advanced industrial and technical community to risk experimenting with some type of neo-rural society in which organic home gardening, bicycles, and total reliance on renewable energy sources are the mainstays.

Much will depend on the future of cities. High density population areas in our culture cannot continue without high-density energy sources, which is one reason why solar panels and wind generators seem totally unable to promise any major on-site contribution to the overall energy needs of a place like Manhattan, regardless of what they might do some day in New Mexico. Yet population dispersion requires energy transport, unless one stipulates a degree of self-sufficiency in energy production that challenges common sense. That brings this discussion back to energy distribution and delivery systems.

Relative decentralization of energy consumers need not affect the concentration of basic energy supply. Coal, the nation's most abundant domestic fuel resource, could still be mined in large enterprises—both for the sake of efficiency and in the interest of confining and controlling adverse environmental effects. If it is burned *in situ*, its energy content could be transformed either into easily moved electricity or, via chemical processing, into gaseous or liquid fuels that might be delivered by pipeline networks. Alternatively, the threatened traffic jam of bulk railroad delivery for coal (especially across the Western United States, where so many grade crossings pose twin problems of delay and vehicular accidents) might be ameliorated by coal slurry pipelines. Such pipelines would carry pulverized coal in a fluid mixture with either water or some burnable liquid. Their implementation has been blocked for years by resistance from rail companies, which are loath to accept competition in the coal-hauling business and which have denied permission for slurries to cross the railroad rights-of-way. That situation could change if the nation as a whole decided it had to.

Existing networks for both gas and electricity are likely to be expanded considerably, even in the light of a conscious effort to curb the rate of growth in energy use. Gas pipelines in this country have already been adapted at least once—in the switch from manufactured gas to natural gas. Because present pipelines already reach the homes and businesses of ultimate consumers, additions to them are a convenient mode of tapping new "unconventional" sources of burnable gas, such as biomass. More gas pipelines will cross international borders too, with diplomatic results that can be conjectured. The U.S. probably will be interconnected with both Canada and Mexico, and success with a gas link between Italy and North Africa should make it only a matter of time before other large bodies of water are crossed. For instance, Japan might connect a gas or liquid fuel umbilical cord to either China or the Soviet Union, and the Japanese choice between those two potential energy partners could affect the delicate balance of world peace.

For technical reasons, gas pipelines of today's design are less likely to play a role in the oft-predicted "hydrogen economy." Gaseous hydrogen is hard to contain; liquid hydrogen must be kept at super-cold temperatures. If hydrogen is generated in large quantities—say, as a means of "storing" energy produced intermittently by renewable sources—it probably would be transported as a compound of some light metal that decomposes under safe, controlled conditions to yield the usable gas. A substantial new support industry might have to evolve.

Electricity is a bit like a marvelously adaptable pipeline. Ultimately, it matters very little whether the generating source is associated with a nuclear reactor, a coal-fired boiler, a dam, a windmill, a solar array, or anything else. The delivery system to end-users is clean, convenient, and to a large extent already in place. Baseload electrical generating capacity could remain concentrated in a relatively small number of locations. The problems of moving large volumes of energy into densely populated areas seem to be easier to overcome with electricity than with most other energy forms. The year 2004 should see superconducting transmission along loss-free lines, presaged by several intermediate improvements along the way.

Nevertheless, the difficulty of electricity storage remains to be overcome. A social solution might be to change time-habits in the use of electricity. The means to do this might range from rate modification to staggered working and living hours that keep things open around the clock. A technological "fix," on the other hand, could be the addition of neighborhood fuel-cell or battery stations that accepted energy whenever it arrived and dispensed it whenever it was needed locally.

With energy and manpower both at a premium, massive capital investments seem to be required in the years ahead if the nation is to maintain productivity and acceptable living standards. This would be the case whether the nation leaned toward a "hard" path, a "soft" path, or something in the middle. Synthetic fuel plants are big ticket items, and so are light rail systems for mass transit. Even "backyard" energy supplies require a huge collective expenditure when one considers the needs of more than a quarter of a billion people. The degree to which investment can and will be made by business, government or individuals raises all sorts of subsidiary questions. Lurking in the background is the possibility of nationalization of the means of production—an idea raised formally by the AFL-CIO's Executive Council in August 1979 in regard to the oil industry.

Unless the nation reverses its field completely, conservation of energy is likely to continue as a social theme—although pressures to keep energy prices down could discourage both conservation and new domestic production. According to historic patterns, new energy sources such as solar heating and cooling will have to make their inroads gradually and in regional patterns. Few of today's energy sources could disappear completely over a time frame of only 30 years—even granting the faster tempo of change in modern life.

Speaking of speed, it is probably a good bet that Americans will reevaluate priorities in that respect—partly because of economic pressures, partly because of social pressures, and partly because instantaneous video communication could make rapid personal travel (as distinguished from the movement of certain goods) less essential. Moving into even more speculative fields, one might envision such anomalies as the institution of freight rockets in the same period as the revival of the relatively low-speed but energy-conserving dirigible. If public apprehension about nuclear energy fades, as it did with earlier applications of technology which had far, far worse safety records, there could be even greater irony. Reactor-propelled hydrofoils larger than the *Queen Elizabeth II* might be taking jet-shunning leisure groups on 24-hour trips across the Atlantic without using a drop of precious oil . . . at the same time even *more* leisurely travelers were making the crossing in enormous 25-knot sailing vessels, whose computer-adjusted rigging and design would be only vaguely reminiscent of the clipper ships.

History suggests that not even the natural resource problems of our energy future can be predicted with complete assurance. Valuable resources have gone unrecognized in the past. Hamilton was astute enough to foresee the long-range importance of coal, but he did not realize the nation's eventual dependence on petroleum—even though American Indians had noticed centuries earlier that surface pools of the stuff would burn. And there are old dictionaries that describe uranium as a "heavy, practically worthless metal."

What are the scarce energy resources of the future, which we might be overlooking now? Will either superconductive technology or lighter-than-air vehicles be desperate for the helium we have been venting from natural gas wells? Will some relatively abundant material like zinc or nickel be precious because it becomes an essential ingredient in chemical battery packs being produced by the billion? Could arable land be a scarce commodity—even in broad America—because new insights into the mechanism of photosynthesis have combined with a new resolve to feed all the hungry of the world at the same time we are cultivating sunflowers as a biomass basis for synthetic fuels? It is hopeless to try to cover all the possibilities in such speculation, yet exercises of this type are not pointless. One of the fundamental lessons from studying energy history is that almost anything *might* happen. Some of the possibilities raised in this chapter may seem a bit bizarre, and they are not actually predictions. But they are the sorts of developments that *could* take place—because all bear certain similarities to past results in comparable situations.

Is a knowledge of energy history helpful in framing energy policy? It could be. At the very least, it commends the wisdom of a balanced view and a non-dogmatic approach.

Either planned or unplanned energy policy resembles a mobile. The desirable goals of policy are like the dangling arms of the mobile, with each weighing against the other. Concentrate on one and you risk disturbing the rest, because they are all interconnected. Balance is the key.

For most of U.S. history we have had no *formal* energy policy, yet five implicit goals have generally been discernible—all interconnected. There is no reason to suspect those same policy elements will not continue to be important between now and 2004.

1) It makes sense to concentrate on the resources close at hand, or easily and reliably accessible. Over a century ago, this forced railroad locomotives to switch from wood to coal. In today's shrunken world, this same principle underlies our focus on domestic supplies.

2) When all other factors are equal, most people prefer a commodity with the lowest price. The "whale oil story" contains a kernel of truth, but a country as big as the U.S. always will use *many* forms of energy because of regional variations in price.

3) Comfort and convenience—including the reliability of supply—are real-life factors an energy planner ignores only at some peril. On a strict basis of the average price-per-unit-of-energy, coal was a better buy than oil during the 1918-1946 period, and coal was plentiful. Yet that is precisely when coal faded from the front of the energy stage. Coal was *not what we wanted*, and most of the reasons for that attitude were related to comfort and convenience.

4) Public health, public safety and the preservation of pleasant surroundings are not new interests. It is foolishly anachronistic to compare precise modern measurements of "air quality" with either the virgin wilderness of the late 1700s or the smoky smelters of the late 1800s, but the principle of protecting people and nature has always been seen as desirable.

5) This is an impatient nation, which has ached to solve problems as quickly as possible. If anything, this tendency has increased. Nevertheless, because of size and the complexity of society, it takes time to effect change. A careful reading of energy history confirms that.

A "solution" to the present U.S. energy problem could be speeded up, and it is true there is no time to waste. Yet, focusing only on the fifth goal, by dictating a solution because of some arbitrary deadline, would upset chances of maintaining reasonable progress toward the other four goals. In fact, all five of the above goals are desirable and always have been. All five have always been unachievable to a certain extent, too. The road to real progress in devising energy policy will take some turns dissatisfying to each group among us . . . yet we must all move along the road *together* or risk chaos. We would be well advised as a nation to appreciate that fact. This may be the best basis for orderly discussion of our various energy options.

SELECTED BIBLIOGRAPHY

Among the scores of publications which served as direct or indirect sources for information and ideas in this brief work, the following are among those which the reader may find most interesting and/or fruitful for further examination:

1. General - Statistical

U. S. Department of Commerce, Bureau of the Census, *Historical Statistics of the United States, Colonial Times to 1970* (Bicentennial Edition, 2 vols.), 93rd Congress, 1st Session, House Document No. 93-78, Washington, D. C., 1975.

Sam H. Schurr, Bruce C. Netschert, Vera F. Eliasberg, Joseph Lerner and Hans H. Landsberg, *Energy in the American Economy, 1850-1975*, published for Resources for the Future, Inc. by the Johns Hopkins University Press, Baltimore, 1960. (The original edition is out of print, but the book in its entirety was reprinted in 1977 by Greenwood Press, Westport, Conn.)

J. Frederic Dewhurst and Associates, *America's Needs and Resources*, The Twentieth Century Fund, New York, 1955.

Sam H. Schurr, Joel Darmstadter, Harry Perry, William Ramsay, and Milton Russell, *Energy in America's Future: The Choices Before Us*, published for Resources for the Future, Inc. by the Johns Hopkins University Press, Baltimore, 1979. (See especially chapters 1, 3 and 4.)

John C. Fisher, *Energy Crises in Perspective*, John Wiley & Sons, New York, 1974.

2. Cultural Context

J. C. Furnas, *The Americans: A Social History of the United States, 1587-1914*, G. P. Putnam's Sons, New York, 1969.

J. C. Furnas, *Great Times: An Informal Social History of the United States, 1914-1929*, G. P. Putnam's Sons, New York, 1974.

Daniel J. Boorstin, *The Americans: The National Experience*, Vintage Books, New York, 1965.

Daniel J. Boorstin, *The Americans: The Democratic Experience*, Vintage Books, New York, 1973.

Peter d'A. Jones, *America's Wealth: The Economic History of an Open Society*, The Macmillan Company, New York, 1963.

Robert William Fogel and Stanley L. Engerman (editors), *The Reinterpretation of American Economic History*, Harper & Row, New York, 1971.

Allan Nevins and Henry Steele Commager, *A Pocket History of the United States*, Washington Square Press, Inc., New York, revised edition, 1967.

Roger Burlingame, *March of the Iron Men*, Grosset and Dunlap, 1938.

Ellis L. Armstrong, *History of Public Works in the United States, 1776-1976*, American Public Works Association, Chicago, 1976.

W. H. G. Armytage, *A Social History of Engineering*, Faber and Faber, London, 1961. (Contains an extensive and useful bibliography).

Nathan Rosenberg, *Technology and American Economic Growth*, Harper & Row, New York, 1972.

Robert L. Breeden (editor), *Those Inventive Americans*, The National Geographic Society, Washington, 1971.

3. Individual Energy Sources

Howard N. Eavenson, *First Century and a Quarter of American Coal Industry*, Pittsburgh, 1942 (privately printed and generally available only in good research libraries, but rich in detail).

Harold F. Williamson and Arnold R. Daum, *The American Petroleum Industry, 1859-1899: The Age of Illumination*, Northwestern University Press, Evanston, 1959.

Harold F. Williamson, Ralph L. Andreano, Arnold R. Daum and Gilbert C. Klose, *The American Petroleum Industry, 1899-1959: The Age of Energy*, Northwestern University Press, Evanston, 1963.

Corbin Allardice and Edward R. Trapnell, *The Atomic Energy Commission*, Praeger Publishers, Inc., New York, 1974.

"The Electric Century, 1874-1974," a special 442-page edition of the McGraw-Hill magazine *Electrical World*, dated June 1, 1974.

"Creating the Electric Age," a special edition of *EPRI Journal*, published by the Electric Power Research Institute, Palo Alto, California, in March 1979.

For still narrower subjects, the author especially commends several series of brief but authoritative pamphlets published by members of the curatorial staff of the National Museum of History and Technology. Some have been published as catalogs of special exhibits, others have appeared as museum bulletins, and still others (covering such topics as "Steam Locomotives," "Spinning and Weaving" and "Horse-Drawn Vehicles," are part of a group published under the general title *In the Smithsonian*. Although their focus is principally technical, they often provide extra insight into the socio-economic-cultural milieu in which specific energy-related devices were introduced or popularized.